DESIGNING FOR SOCIAL JUSTICE

Exploring the intersection of design research and community engagement, this book highlights the ways in which design and design theories can be used to address social justice issues and promote positive change in communities.

Contributors illuminate the theoretical, ethical, and pedagogical dimensions of design-driven methods in community-engaged projects, exploring their potential to address critical social justice issues such as ethnic and racial justice, gender equality, disability justice, cultural diversity, equity, and environmental justice. Chapters examine various aspects of community-engaged practices, including the use of design theories to fuel social justice work in community partnerships, ethical issues surrounding the use of multimodal resources and new media technologies, and pedagogies for promoting social change. Addressing the opportunities and challenges of design and design methods in community engagement, this collection offers suggestions for promoting social justice through technical and professional communication activities and pedagogies.

Investigating the design of community-engaged projects from a critical standpoint, this book will appeal to scholars and students in the fields of Technical and Professional Communication, Writing and Composition Studies, and Rhetoric. It will also be of interest to administrators, community partners, and professionals working in service-learning contexts.

Jialei Jiang is a Teaching Assistant Professor of Composition in the Department of English at the University of Pittsburgh, USA.

Jason C. K. Tham is Associate Professor of Technical Communication and Rhetoric and Assistant Chair of the English Department at Texas Tech University, USA. His recent works include *UX Writing* (Routledge, 2023), *Designing Technical and Professional Communication* (Routledge, 2021), and *Design Thinking in Technical Communication* (Routledge, 2021).

ATTW Series in Technical and Professional Communication

Michele Simmons and Lehua Ledbetter, Series Editors

For additional information on this series please visit www.routledge.com/ ATTW-Series-in-Technical-and-Professional-Communication/book-series/ATTW, and for information on other Routledge titles visit www.routledge.com.

DESIGNING FOR SOCIAL JUSTICE

Community-Engaged Approaches in Technical and Professional Communication

Edited by Jialei Jiang and Jason C. K. Tham

Routledge
Taylor & Francis Group

NEW YORK AND LONDON

Designed cover image: Original artwork by Jialei Jiang

First published 2025
by Routledge
605 Third Avenue, New York, NY 10158

and by Routledge
4 Park Square, Milton Park, Abingdon, Oxon, OX14 4RN

Routledge is an imprint of the Taylor & Francis Group, an informa business

ISBN: 978-1-032-74592-3 (hbk)
ISBN: 978-1-032-74591-6 (pbk)
ISBN: 978-1-003-46999-5 (ebk)

DOI: 10.4324/9781003469995

Typeset in Sabon
by KnowledgeWorks Global Ltd.

We dedicate this project to those who are relentlessly engaged in community-driven efforts to create the worlds we need.

CONTENTS

FIGURES AND TABLES

Figures

Tables

CONTRIBUTORS

Jamal-Jared Alexander is an Assistant Professor of Technical and Professional Communication at the University of Tennessee, Knoxville. His research agenda aims to create dedicated spaces and equitable access/opportunities for marginalized communities in academic and industry settings. His collaborative and interdisciplinary scholarship has won national and international awards, including awards for best article in rhetoric of science and technology, and an early career research award. His research has informed recruitment and retention initiatives at different institutions, focusing on inclusive excellence at the department, college, and university levels.

Dorcas Anabire is a third-year Ph.D. student in Technical Communication and Rhetoric and a Presidential Doctoral Research Fellow at Utah State University. Her research focuses on ways design can be used to address social justice issues in workplace and academic settings. Specifically, she is interested in how specific design practices can improve usability, user experience, and accessibility, especially in cross-cultural contexts and marginalized communities, and how they intersect with social justice. She also researches inclusive recruitment practices for recruiting minoritized students, specifically international students in higher education.

Felicita Arzu-Carmichael is an associate professor of writing and rhetoric at Oakland University, where she teaches courses in race, social justice, and professional writing, technical writing, and first-year writing. At Oakland, Felicita also serves as associate director of the First-Year Writing Program. She is an associate editor of *College English*, the flagship journal of the college section of NCTE. Felicita's scholarly interests focus on online literacy, first-year writing, and race, social justice, and inclusion. She is a recipient of the 2023-2024 CCCC Emergent Research Award, and her work has appeared in *Technical Communication Quarterly*, *Writing Program Administration*, *Composition Studies*, and other journals. Felicita is a native of Belize. In her home country, she is an executive member of the National Garifuna Council (NGC), Orange Walk branch. The NGC is a non-governmental organization committed to preserving, strengthening, and advancing the rights and culture of the Garifuna, an Indigenous Afro-Caribbean people.

Sweta Baniya is an assistant professor of rhetoric and professional and technical writing and an affiliate faculty of Women and Gender Studies at Virginia Polytechnic Institute and State University. Her book *Transnational Assemblages: Social Justice and Crisis Communication During Disaster* explores transnational activism in the April 2015 Nepal Earthquake and 2017 Hurricane Maria in Puerto Rico. Dr. Baniya's scholarship has appeared in *Technical Communications Quarterly*, *Technical Communications*, *Enculturation*, *Journal of Business & Technical Communication*, and *Journal of Technological Studies*, among others.

Antonio Byrd is an assistant professor of English at the University of Missouri-Kansas City. He studies how the legacies of literacy for racial liberation carry forward into contemporary digital literacy practices of Black communities. His most recent work focused on computer programming as literacy and how Black adults leverage its features for personal well-being. Antonio's work has appeared in *Literacy in Composition Studies*, *Technical Communication Quarterly*, *College Composition and Communication*, and two edited collections. His book *From Pipeline to Black Coding Ecosystems: How Black Adults Use Code Bootcamps for Liberation* is forthcoming with The WAC Clearinghouse/University Press of Colorado.

Erin Brock Carlson (she/her) is an assistant professor in the Department of English at West Virginia University, where she teaches undergraduate and graduate courses in technical and professional writing, research methods, and digital humanities. Her current research addresses community organizing, place-based development, and collaborative knowledge-making through participatory visual methodologies. Committed to public-facing and public-serving humanities work, Erin strives to incorporate community-based projects into her courses and prioritizes research collaborations with organizations and communities in the Appalachian region. Her work has appeared in *Technical Communication Quarterly, Journal of Business and Technical Communication, Communication Design Quarterly,* and more.

Ellen Cecil-Lemkin (she/her) is an assistant teaching professor at the University of Wisconsin-Madison, where she helps direct the Writing Center. Her research focuses on disability, accessibility, and collaboration. Her scholarship has been published in *College Composition and Communication, Teaching English in the Two-Year College,* and several edited collections. When she's not working, she can be found chasing after her two young children.

Jared S. Colton is the Associate Dean for Undergraduate Studies in the College of Humanities and Social Sciences and an Associate Professor of Technical Communication and Rhetoric in the Department of English at Utah State University. His research addresses the intersections of rhetorical theory, ethics, and digital technologies in technical communication and related fields. He has work published in *Rhetoric Review, Journal of Business and Technical Communication, Technical Communication Quarterly, Computers and Composition,* and other journals. His co-authored book *Rhetoric, Technology, and the Virtues* was published by the University Press of Colorado.

Ahlan Filstrup is an undergraduate English student at Kennesaw State University. She has worked closely with the Atlanta Student Movement Project as a student researcher. In this role, she designed several first-year composition curricular assignments centering on civil rights and social justice topics. She has also co-written a book chapter on Artificial Intelligence in the first-year classroom and has presented research on social justice education at the Southern Regional Honors Conference, the Linguistics Symposium of North Georgia, and the National Conference of Feminisms and Rhetorics.

Tobechukwu Precious Friday is a Distinguished Fellow of the Institute of Leadership, Manpower & Management Development, Co-founder & Executive Director of Igbo Wikimedia Organization and a Wikimedian in Residence at Wikitongues. She holds a B.A in Foreign Languages and Literary Studies from the University of Nigeria, Nsukka, a Masters in Project Management from Rome Business School as well as a Doctorate in Leadership and Business Administration (Honoris Causa). She is a member of the Wikimedia Foundation's Language Committee and a Board Member at Wiki in Africa. She speaks English, Igbo French, and German.

Laura Gonzales is President of the Association of Teachers of Technical Writing. She is the award-winning author of *Designing Multilingual*

Experiences in Technical Communication and *Sites of Translation: What Multilinguals Can Teach Us About Digital Writing and Rhetoric*. Dr. Gonzales advocates for the importance of linguistic diversity in community, technological, and academic contexts.

Carrie Grant is an assistant professor in the Department of English at Towson University, where she co-directs the Grantwriting in Valued Environments (GIVE) community writing program. Her research interests include nonprofit writing, community engagement methodologies, feminist digital literacies, and social justice in technical communication. Her work has appeared in *Communication Design Quarterly, Computers and Composition, IEEE Transactions on Professional Communication*, and *Open Words: Access and English Studies*.

Serenity Hill is a Limited-Term First-Year Composition Instructor at Kennesaw State University. She holds an M.A. in Professional Writing with certifications in Professional Editing and Publishing and Creative Writing, from Kennesaw State University. She has experience as a literary magazine editor and has published both poems and film reviews. In addition to her creative work, her research focuses on using archival research to investigate and educate others on the untold stories of minorities, specifically African American and women writers, editors, and publishers. She has presented this work at the Feminisms and Rhetorics Conference and at the Conference on College Composition and Communication. Additionally, she was named the Exceptional MAPW Capstone Project for the 2023-2024 Academic Year award recipient for her research highlighting the contributions of Black women writers, editors, and publishers during the Civil Rights Movement. The goal of her work is to educate students on their history which she has begun to do by creating and implementing a curriculum that advocates for representation, acknowledgment, and discussion of social justice in the classroom and the publishing industry.

Steve Holmes is an associate professor of technical communication and rhetoric at Texas Tech University. His research areas include ethics, rhetoric, social justice, and digital technology. Steve has published two books (*Procedural Habits*; *Rhetoric, Technology and the Virtues* with Jared S. Colton) and an edited collection (*Reprogrammable Rhetoric* with Michael J. Faris). His recent research explores the ethics of content moderation and banning as a form of ethical infrastructure through various forthcoming publications with Danielle Feldman Karr, Rachael Jordan, Jared S. Colton, and Josephine Walwema.

Jialei Jiang is a teaching assistant professor in the Department of English at the University of Pittsburgh. She is interested in exploring issues related to multimodal composition, critical AI literacy, and socially just pedagogy. Her works have appeared in *College Composition and Communication, Computers and Composition, Critical Studies in Education, the Journal of Technical Writing and Communication, Postdigital Science and Education, TESOL Quarterly*, and edited collections. Currently, she serves as an assistant editor of the journal *Kairos: A Journal of Rhetoric, Technology, and Pedagogy*. At the University of Pittsburgh, she instructs courses in composing digital media, technical writing, and multilingual composition. Her experience with multimodal design has extended her ability to explore innovative theoretical and pedagogical approaches surrounding the integration of emerging technologies, such as generative AI, into multimodal composition.

Aimée Knight is a Professor of Communication and Media Studies at Saint Joseph's University. She is the author of *Community is the Way: Engaged Writing and Designing for Transformative Change* (2022) and the founder of the Beautiful Social Research Collaborative. Her work centers on community-led design and social equity, with a focus on dismantling traditional power dynamics in research and higher education. Aimée's research and teaching explore the intersections of community engagement, social justice, and media, driving real-world impact through over 130 local and international engaged writing partnerships.

Danielle Marie Koepke is a Teaching Assistant Professor at Marquette University where she teaches classes such as Business Writing, Health Science Writing, and Writing for Social Change. Her research areas include cultural rhetorics, reproductive justice, and digital design; these areas influence her practices of care as a teacher and as a community-engaged researcher. In her current book project, she is studying the strategies for and impacts of sharing stories of reproductive [in]justice on social media since the *Dobbs* decision. Her work can be found in publications such as *Composition Studies* and in edited book collections including *Digital Literacies for Human Connection* (NCTE), *Rhetorica Rising, Practicing Digital Activisms: On Rhetoric, Writing, and Technical Communication's Social Justice Obligations*, and more.

Jeanne Beatrix Law is a Professor of English at Kennesaw State University. Her research specialties include multimodal languaging, digital histography, and generative AI technologies for writers. Her public scholarship includes scaling historical rhetorics for diverse audiences and emergent modalities. Jeanne is the co-author of *The Writer's Loop: A Guide to College Writing* and was a founding author of Andrea Lunsford's *Multimodal Mondays* blog series. She has authored chapters on information literacy in edited collections from Purdue press, SIU press, and Routledge. Her work is also regularly featured in print and digital public media, including *The Chronicle of Higher Education*. She is the lead researcher for the nationally recognized #ATLStudentMovement oral history project and has authored eight courses on Coursera on generative AI use, featuring her Rhetorical Prompt Engineering and Four Qualifiers for Ethical Outputs frameworks. Jeanne also serves as a faculty mentor for the AAC&U's AI Pedagogy Institute. She serves as the Writing Program Administrator for Kennesaw State's general education program and has been called in as an AI usage expert by the University System of Georgia on numerous occasions. She has three chapters on AI use cases accepted by Computers & Composition and Routledge. She has presented and published numerous times since 2022 for professional, public, and academic audiences on the ethical use of generative AI.

Robin Lewy is Co-Founder and Director of the Rural Women's Health Project, the only health justice organization in North Florida. Robin has worked for 35 years in social justice, health education, and community mobilization in the United States and Central America. In her role, she oversees advocacy and direct service programming that serves Latinx immigrants, farmworkers, and rural communities. Robin is a regional trainer on health inequity in rural communities, Promotor de Salud programming, and the use of testimonial media. She is the Community Liaison for the Gainesville Immigrant Neighbor Inclusion Initiative which has brought together 60 members into a Steering Committee to develop a Blueprint for immigrant integration in the city and county. She holds an MA in Social Development and Popular Theatre, the University of Nebraska-Omaha.

Sarah Moon is a Visiting Assistant Professor in the Department of Professional and Public Writing at Roger Williams University. She is the founder and director of the community writing and performance project Write Your Roots. Her scholarly work has been published in *Pedagogy*, *Community Literacy Journal*, *Literacy in Composition Studies,* and *Center for Sustainable Practices in the Arts Quarterly*.

Jennifer Nish is an associate professor of rhetoric and writing at Michigan Technological University. She has previously held faculty positions at Texas Tech University and the American University of Beirut. Her research and teaching focus on writing studies, transnational feminism, digital media, disability, and activism. Her published work includes *Activist Literacies: Transnational Feminisms and Social Media Rhetorics* (2022); articles published in *Peitho* and *College Composition and Communication*, and chapters published in textbooks and scholarly collections. She is currently working on a collaboratively authored first-year composition textbook, a co-edited collection on transnational feminist rhetoric, and a series of projects about gender-, sexuality-, and disability-related activism.

Therese I. Pennell is an assistant professor and director of the writing center at Prairie View A&M University in the Languages and Communication Department. She is originally from Belize.

Lisa L. Phillips is an assistant professor of English in the technical communication and rhetoric program at Texas Tech University. She conducts research and teaches graduate and undergraduate courses on environmental rhetoric, intersectional feminism, social justice, technical communication, and sensory rhetorics. Her forthcoming book (2025) *Olfactory Rhetoric: Sniffing Out Environmental Problems* expands on the ideas presented in this chapter. Her co-edited collection *Grassroots Activisms: Public Rhetorics in Localized Contexts* published with The Ohio State University's "Intersectional Rhetorics" series is available as open access: https://openresearchlibrary.org/content/fdd451a7-fcbf-4016-90cb-1dd47daf9c09.

Katrina Powell is Professor of Rhetoric and Writing and Founding Director of the Center for Refugee, Migrant, and Displacement Studies at Virginia Tech. Her research focuses on displacement narratives, human rights, ethics in research methodology, and oral history. Her books include *The Anguish of Displacement* (2007), *Identity and Power in Narratives of Displacement* (Routledge, 2015), and *Beginning Again: Stories of Movement and Migration in Appalachia* (2024). She also serves as co-director of Monuments Across Appalachian Virginia, a $3 million Mellon Foundation grant to create monuments that focus on migration and mobility in Appalachian Virginia.

Ahoo Salem is the Executive Director of Blue Ridge Literacy (BRL), a non-profit organization based in the Roanoke Main Library Building that provides English and digital literacy and citizenship preparation services to foreign-born adults in the Roanoke Valley. As a sociologist with a deep passion for integration initiatives, Ahoo is particularly interested in how differential access to resources shapes and influences the everyday life experiences of immigrants in diverse host settings. Ahoo currently serves on the Emergency Management Equity Working Group (EMEWG), which ensures that emergency management programs address the needs of vulnerable populations. She previously served on the Virginia Office of New Americans Advisory Board (2020-2024), where she co-led efforts to support the linguistic and cultural integration of New Americans in Virginia. Originally from Iran, Ahoo holds a Ph.D. in Sociology from Università degli Studi di Milano (Milan University) in Milan, Italy.

Layla Scott is a senior majoring in computer science in the College of Engineering at Virginia Tech. She is interested in human-computer interaction (HCI) and ways to improve user interfaces. During her undergraduate studies, Scott developed a mobile application via ACLS grant to help immigrants study for their citizenship civics test.

Valentina Sierra-Niño is a Colombian designer who explores the potential of design to help us co-exist in more empathic ways within our communities. As a Fulbright scholar, she earned her master's degree in Design and Visual Communication from the University of Florida. Valentina works as a professor at Universidad de los Andes, in Bogota, Colombia, where she teaches courses on designing with communities and design research methodologies.

Kristen Tcherneshoff is the Programs Director at Wikitongues. Kristen received a BFA in photography from the University of Alabama and holds an MA in African studies and linguistics from Helsinki University. With a background in law and disability rights, her interests lie in language activism and sign linguistics. She speaks English and Finnish.

Jason C. K. Tham is an associate professor of technical communication and rhetoric at Texas Tech University and assistant chair of the English department. His Routledge titles include *Design Thinking in Technical Communication* (2021), *Designing Technical and Professional Communication* (with Deborah Andrews, 2021), and *UX Writing* (with Tharon Howard and Gustav Verhulsdonck, 2024). He is editor of the collections *Keywords in Design Thinking* (2022) and *Keywords in Making* (2024). He serves as chief editor of *Computers and Composition* and vice president of the Council for Programs in Technical and Scientific Communication. Previously, he was an inaugural Faculty Fellow of engaged scholarship, sponsored by TTU Outreach & Engagement and Office of the Provost.

Daniel Bogre Udell is the Co-Founder and Executive Director at Wikitongues, where he supports the leaders of new and early-stage language revitalization projects. He began participating in global online initiatives in 2011, as a contributor to the nonprofit news organization Global Voices. There, he launched a Catalan-language edition, which he co-edited until 2015. He is a 2019 TED Resident and a recipient of the 2021 J.M.K. Innovation Prize. Daniel holds a BFA in design and technology from Parsons School of Design, a BA in History from Eugene Lang College, and an MA in historical studies from The New School for Social Research. He speaks English, Spanish, Catalan and Portuguese.

Rebecca Walton is the executive Associate Dean for the College of Humanities and Social Sciences at Utah State University and a Professor of Technical Communication and Rhetoric in the Department of English. Walton researches how people intervene for justice in their workplaces. Her co-authored scholarship has won multiple national awards, including awards for best book, best theory article, and best empirical research article. Her research has informed implicit bias training, policy revision, and curriculum development at multiple universities, and her leadership experience includes external program review, journal editorship, and program redesign at the undergraduate and graduate levels.

Josephine Walwema is with the University of Washington, Seattle where she coordinates the program in technical and professional writing. Walwema serves as associate editor of *IEEE Transactions on Professional Communication*. Her research, which centers on issues of access and social justice in technical and professional communication, has been published in *Technical Communication Quarterly*, *Technical Communication*, *Connexions*, and *IEEE*. She is a recipient of the 2017 Nell Ann Pickett Award for best article in *Technical Communication Quarterly* and has twice won the CCCCs Best Article in Reporting Historical Research or Textual Studies in Technical and Scientific Communication.

Xiaobo Wang is an Assistant Professor of English and Assistant Director of MA in Technical Communication at Sam Houston State University. Her research focuses on the intersections of global communication design, social justice, virtue ethics and contemporary technologies, feminist rhetoric, comparative rhetoric, and most recently, global health and risk communication. She has published in journals such as *Technical Communication*, *Journal of Business and Technical Communication*, and *Communication Design Quarterly*. She serves as an Associate Editor of *IEEE Transactions on Professional Communication* and co-chair of the Asian/Asian American Caucus, Conference on College Composition and Communication.

Margaret Webb, a Ph.D. student in Engineering Education at Virginia Tech, aims to revolutionize engineering education to foster interdisciplinarity in disaster resilience and displacement research. Inspired by her Hurricane Katrina experience, Webb explores how engineers inadvertently impact communities through disaster resilience and climate change work. She develops interdisciplinary learning spaces to encourage holistic thinking about engineering's consequences. Webb integrates social sciences, ecology, and ethics to equip engineers with tools to assess their work's full impact, develop inclusive resilience strategies, and engage with stakeholders. With a background in Mechanical and Civil Engineering, subsea engineering, and STEM teaching, Webb bridges technical expertise and societal implications to cultivate ethically-minded engineers for complex global challenges.

Liping Yang is a writer, teacher, and scholar with a deep interest in digital rhetoric, critical AI literacy, human-machine communication, and the digital humanities. Currently, she is a PhD candidate in Rhetoric and Composition at Georgia State University (GSU), where she also serves as the director of the First-Year Writing Program and as a graduate associate in the Writing Across Curriculum (WAC) program. She has published translations, academic journal articles, and book chapters, and she also presents her work at various academic conferences focusing on college composition, technical communication, and the English language. Besides, she also holds leadership roles in national organizations such as AWAC, STC, and NeMLA. At GSU, she instructs courses such as first-year composition, digital writing, technical communication, etc.

ACKNOWLEDGMENTS

First, we would like to acknowledge the Sweetland Digital Rhetoric Collaborative for bringing the two of us together via its graduate student fellowship in 2018–2019. Through digital projects such as blog carnivals and wiki writing, we found shared interests in our research and teaching. As cited in the Editors' Introduction of this book, our 2019 blog carnival, titled "Multimodal Design and Social Advocacy," was the genesis of this collection.

Second, we are grateful to the Emergent Researcher Award granted to us in 2020 by the National Council of Teachers of English (NCTE) and Conference on College Composition and Communication (CCCC). The award and its seed funds served as a catalyst to our collaborative scholarship on community-engaged pedagogy. We also thank Natasha Jones, who was assigned as a mentor to us, for her advice on the design of our proposed study. Of course, we are indebted to our research partners, participants, and students for their involvement in the project.

Next, we appreciate the knowledge-sharing spaces made available by professional communities such as the CCCC, Association of Teachers of Technical Writing (ATTW), Computers and Writing, and other conferences where we got to share the findings of our project. If you have attended any of our previous presentations and provided comments, please know that you also helped shape this book.

For our project to take the form of a book, the press plays a vital role. We must give a shout-out to the editors, editorial assistants, and production team at Routledge, including Alexandra de Brauw, Sean Daly, and Zoya Gayle, who made this publication a possibility. We also thank the

ATTW Book Series co-editors, Michele Simmons and Lehua Ledbetter, for supporting this title.

Finally, we thank the contributors to this collection for sharing their experience and knowledge. These authors are generous in the reporting of their respective projects, including stories of success and challenges, methods and strategies for designing community-engaged projects, and recommendations for teaching students to participate in these projects.

Jialei Jiang:

I am thrilled to have had the opportunity to work alongside Jason. His brilliance and collaborative spirit have been invaluable. I deeply appreciate his initiative in serving as the corresponding editor for this book and his trust in sharing the co-leadership with me.

I extend my heartfelt thanks to Matthew Vetter for his insightful feedback on the early versions of our project proposal. His constructive critiques and suggestions significantly enhanced the quality of this book.

To my family, especially my husband Joseph Noel and our child Leo Noel, your unwavering love has transformed the writing process into a joyous experience. Thank you for your constant support and encouragement.

Jason Tham:

Writing continues to never be a solo effort in my career. I am most delighted to share this writing space with Jialei. She is one of the smartest scholars I know who is also a fantastic collaborator. I am thankful for her idea to propose this book project and for believing in me enough to share the co-pilot responsibility.

I am also grateful to my colleagues, students, and administrators at Texas Tech University for recognizing the significance of this project and allowing me to dedicate time to completing it.

To my family, including my partner Kamm and our needy furbabies, I am thankful for the love and flexibility that made writing this book an enjoyable process.

EDITORS' INTRODUCTION

Social Justice and Multimodal Design in Ethical Community Engagement

Motivation

Over the last 20 years, community engagement has gained considerable attention from scholars, teachers, and practitioners in the field of technical and professional communication (TPC). TPC scholars design projects in collaboration with community partners to create community-engaged research, teaching, and learning. In community-based research, scholars have examined the influence of community participation in civic projects (Carlson, 2020; Simmons, 2007), researcher roles in the community (Blythe et al., 2008; Edenfield & Ledbetter, 2019), technological impact (Grabill, 2007), and most recently, justice-oriented approaches and efforts in engaged scholarship (Amidon et al., 2023). Scholars interested in the benefits and challenges of community-engaged learning have studied students' reflections on their service-learning experience (Dubinsky & Carpenter, 2004; Matthews & Zimmerman, 1999; Patterson, 2015), citizenship and social awareness (Agboka, 2014; Mara, 2006; Sapp & Crabtree, 2002), as well as technology use (Cagle, 2017; Haas, 2012; Turnley, 2007). Outside the classroom, community engagement augments scholarly and creative activities to propel applied research, policy development, technology transfer, and volunteerism, to name a few.

As we believe research on service-learning should move in pace with current issues and field focus, a major interest to our project is the emergent social justice lens toward TPC practice and relatedly its pedagogy. Since Jones's (2016) influential argument for the promotion of human agency through collaborative and respectful actions, and the award-winning book

DOI: 10.4324/9781003469995-1

Technical Communication After the Social Justice Turn (Walton et al., 2019), major conversations in TPC scholarship have considered the way a social justice lens influences TPC traditions. Admittedly, community engagement can benefit from a perspective that attends to diversity, equity, and inclusion issues. Colton and Holmes (2018) gave TPC scholars from a rhetorical background even more reasons to focus on ethical and tactical justice efforts through virtue theories. Yet, research on the integration of social justice in community engagement remains scarce in TPC scholarship and under-valued by institutions (Amidon et al., 2023). Hence, this project seeks to investigate the design of community-engaged projects from a critical standpoint. Informed by Costanza-Chock's (2020) critical discussions of design, we apply the notion of "design advocacy" to amplify the concept of multimodality within design-driven efforts to promote social advocacy (Tham & Jiang, 2022). By "design-driven," we aim to highlight the way design activities and philosophies can lead advocacy projects in multimodal, interactive ways.

The design of community engagement often manifests in material forms. Haas (1996) has particularly pointed out the material dimensions of literacy and writing, with the term "material" referring to anything that possesses mass or matter, and which uses physical space. For multimodality theory, this includes any tools or resources that cross between the practitioner's artifacts. The connections between materials, composers, and literacy knowledge in the composing environment are often mapped onto the socio-material conditions of learning as a way of problematizing their relations to the wider societal issues. With the growing use of digital composing technologies in the last decade, materiality as a concept for understanding writers' and communicators' rhetorical intentions has been treated with attention toward the affordances (and limitations) of the digital. For example, Spilka (2009) illuminated digital technology's impact on workplace practices and TPC field's trajectory. Armfield et al. (2012) explored how TPC teachers can be more "digitally savvy" given the increasing demands for high-tech instruction. More recently, Tham (2021) joined a growing group of scholar-teachers who recommend a "critical making" manifest in TPC education to give students the necessary opportunities to address social problems with design tools that can turn ideas into reality (Bay et al., 2018; Breaux, 2017; Cooke et al., 2020; Pellegrini, 2021; Wible, 2020). These material-based, contextually influenced approaches to composing and multimodality emphasize the active and dynamic role of tangible things, and the vitality of their interplay with community-engaged projects.

Despite the overlap between design research and community engagement, only recently have there been studies exploring their connection (Simmons & Amidon, 2019). As the significance of multimodal design continues to grow, it is crucial for TPC researchers and teachers to provide theoretical frameworks and pedagogical lenses that can shed light on the productive uses of multimodal resources and new media technologies, both within academia and in broader public contexts. Given the vital importance of this topic, there has only been one book-length text, so far, on the use of emerging media in community-engaged writing—Knight's (2022) *Community Is the Way: Engaged Writing and Designing for Transformative Change*, which introduces design using new media technologies and multimodal resources. While this work is important, it primarily focuses on one researcher's perspective on design. The field would benefit from a more comprehensive framework of design theory and TPC pedagogy in community engagement. Therefore, to better meet the theoretical and pedagogical needs of the discipline, our edited collection aims to approach multimodal design approaches to community engagement more broadly. This collection builds on the significant research of scholars leveraging the potential of community engagement in recent years (Agboka & Matveeva, 2018; Baniya et al., 2022; Knight, 2022; Rivera & Gonzales, 2021; Shah, 2020; Walton & Hopton, 2018). Extending these earlier conversations, we seek to explore the intersection of multimodal design and community engagement for social justice. To do so, this book covers three main areas of exploration, including design theories and ethics in designing for social justice, community-engaged design efforts in action, and pedagogical exemplars of multimodal design for social justice. In addition to its relevance to researchers, instructors, and students in the field of TPC, this book also appeals to professionals in service-learning contexts, such as grant writers, technical writers, and designers working for government and nonprofit organizations.

Key Definitions

The Carnegie Foundation for the Advancement of Teaching (2006), which sponsors the assigning of institutional designations for elective community engagement classifications, described "community engagement" as "the collaboration between institutions of higher education and their larger communities (local, regional/state, national, global) for the mutually beneficial exchange of knowledge and resources in a context of partnership and reciprocity" (n.p.). Accordingly, "community-engaged learning" is an educational approach that focuses on collaboration among teachers, students, and community partners. In service-learning courses, students use writing to explore and engage with the community, while also reflecting on their experiences. Recently, this approach has gained popularity among scholars and educators as a valid teaching method. Experts in writing studies and TPC research advocate for service-learning as a socially responsible and community-focused strategy (Baca, 2012; Deans et al., 2010; Garza, 2013; Holmes, 2016). Deans et al. (2010) highlighted that a key goal of service-learning is to balance community outreach with academic inquiry, ensuring that the course objectives align with community needs. Effective community-engaged courses involve more than just bringing students into community spaces. Garza (2013) emphasized that service-learning in writing classes should be viewed as an opportunity for students to learn through real projects with community partners, rather than merely a means to an end.

We are interested in how community projects are created in different modes and modalities. The term "multimodality" (Kalantzis & Cope, 2012; Kress, 2010; van Leeuwen, 2005) refers to the various modalities—such as printed words, images (both still and moving), sound, speech, music, and color—that authors combine when creating texts. Building on this understanding, Takayoshi and Selfe (2007), in "Thinking about Multimodality," described multimodal texts as documents found in digital environments that employ multiple modalities to convey meaning. To understand multimodal affordances, defined by Kress (2010) as "the question of potentials and limitations of a mode" (p. 192), we look to ongoing discussions among writing and literacy scholars about multimodal composition. In *Multiliteracies for a Digital Age*, Selber (2009) identified three types of technology literacies: functional, critical, and rhetorical. Selber, drawing on the New London Group's concept of "multiliteracies," advises writing instructors and program administrators on developing comprehensive computer-supported composition programs that emphasize these literacies. Shipka (2009), however, viewed multimodal composition as extending beyond the digital-versus-print dichotomy. Shipka explained that her approach is distinct from those who teach "multigenre" or new media texts because her students often create three-dimensional texts and live performances rather than traditional paper or electronic works (as cited in Dark & Baker, 2015, p. 75). Similarly, Palmeri (2013) contended that equating multimodality solely with digital media misrepresents the true complexity of multimodal experiences for students.

This edited volume aims to examine and critique the relationship between design and community engagement at the intersections of justice and multimodal innovation. While "design" can be an ambiguous term that encompasses many areas of practice and deliverables, we consider it to be a means for activating change, understanding needs, challenging the status quo, and manifesting tangible and desirable differences. This understanding of design can inform the approaches TPC practitioners take to build and maintain community-engaged projects. In the next section, we explore the intersection of design and the ethics of community engagement.

Design and the Ethics of Community Engagement

Whereas the core principles for facilitating successful community engagement often include approaches for trust building, collaboration, leadership, and shared innovation (Moore, 2017), we posit that a design-driven mindset is necessary for justice-oriented engagement. Noted earlier, we understand design to be more than a knack or skill for creating style-specific products (e.g., documents, structure, systems), but a social activity involving critical consideration of the rhetorical functions of forms and delivery, as well as the intentions of design (Eyman & Ball, 2014). For community-engaged projects, design is also the process of determining who gets to be included in the project through participation and representation in sampling, research, delivery, assessment, and other methods (Shivers-McNair, 2017). Thus, the design of community engagement requires attention toward ethics as it affects the mindset and practice of scholars and community members.

Like numerous contributors in this collection, we subscribe to "design thinking" as a human-centered and ecological framework for constructing justice-oriented community partnerships (Tham, 2021). Design thinking is founded on a user-centered design approach to technology solution, which necessitates an empirical process to understanding user requirements, attitudinal and behavioral patterns, contextual issues, and other technical and social needs. The foundation of design thinking is empathy toward people, particularly those who are impacted by complex issues that are often difficult to define. Designers are encouraged to take a collaborative and community-based approach to innovate solutions, build and test prototypes, and iterate their ideas based on feedback and careful evaluation. Despite its widespread adoption, however, design thinking needs an ethical intervention to help practitioners mitigate biases about the social, historical, and political conditions of the human experience (Kimbell, 2011, 2012).

An ethical emphasis in a designerly mindset today requires dedication to advocacy, and advocacy through community-engaged activities does not always come naturally or uncontested (Bay, 2022; Zdenek, 2020). Tensions exist in negotiations of power, agency, compliance, and cultural dynamics in institutional and corporate contexts (Bennett & Hannah, 2021). For TPC scholars and teachers, ethical theories are nevertheless helpful beacons that guide decision-making in community engagement. Yet, there remains a critical need to unpack and inspect these theories *in situ*—to identify the attributes of selected ethics for design and study its implications within applied contexts. Moreover, to actualize justice through design, we are called to scrutinize ethical actions from multiple dimensions (Cox, 2019; Edenfield, 2017; Itchuaqiyaq & Matheson, 2021). In fact, we find our initial questions from a blog carnival in 2019, the provenance of this project, to still be relevant here (Jiang & Tham, 2019):

1 How might we further redefine or re-contextualize the rhetoric of design to include considerations of ethics, advocacy, and marginalization towards issues of social, material, and technological equity?
2 How does design advocacy allow us to reframe and rethink the relevant theories and practices within the individual disciplines of (digital) rhetoric, multimodal composition, and technical communication?
3 In what ways can we sustain design advocacy across the borders and boundaries of these fields of study? How will design advocacy disrupt disciplinary boundaries and/or carve out new spaces of inquiry?
4 What challenges or difficulties may distract us from leveraging the full potential of multimodal design in fostering social advocacy? What are ways to address and overcome these challenges?

Underlining these inquiries was the intent to extend the work of TPC into ethical engagement with community issues through design. Since asking these questions, we have found the social justice lens to be most suitable for situating community-engaged design projects in the TPC landscape.

Social Justice as a Design Concern

Fortunately, our field's scholarship has increasingly emphasized social justice, reflecting a broader commitment to equity and inclusivity in TPC (Leydens, 2012; Walton & Agboka, 2021). This focus involves critically examining how communication practices can perpetuate or challenge systemic inequalities and striving to create more accessible, ethical, and responsive communication strategies. Scholars are exploring ways to integrate social justice principles into TPC curricula, emphasizing the importance of culturally aware and inclusive design. It includes developing frameworks for ethical user experience design (Acharya, 2022; Tham & Grace, 2024), advocating for underrepresented voices (Balghare, 2022), and addressing the societal impacts of technological advancements (Sackey, 2018). By foregrounding social justice, TPC design scholarship aims to produce professionals who are not only skilled communicators but also conscientious advocates for social change.

We believe social change is made possible by design through multimodal means. As indicated before, multimodal projects offer opportunities to develop students' critical and rhetorical skills while enabling instructors to implement meaningful pedagogies addressing social issues. With advances in writing and communication technologies and a growing interest among educators in fostering community-based learning, there's an increase in multimodal writing projects aligned with social advocacy both inside and outside the classroom (Agboka & Matveeva, 2018; Jiang, 2024a, 2024b; Jiang & Tham, 2019, 2023, 2024; Tham & Jiang, 2022, 2024). As well, scholars have investigated the design of digital and multimodal texts as a form of critical design pedagogies (Acharya, 2022; Arola, 2018; Sheridan et al., 2012; Wysocki & Lynch, 2018). This project is a continuation of such investigations.

In *Design Justice: Community-Led Practices to Build the Worlds We Need*, Costanza-Chock (2020) advocated for individuals to engage in critiques of design, broadly conceived as the design of images, software, and technical systems, that contribute to the reinforcement of systemic oppression against women and minorities. Using examples of technologies that range from airport body scanners to Twitter, Costanza-Chock stressed the importance of exploring the social and communal dimensions of design as an important ingredient that could help build design justice for marginalized communities. Extending Costanza-Chock's critical discussions of design, as well as other theorists on the social implications of technology and innovation (Benjamin, 2019; Duin & Pedersen, 2023; Eubanks, 2012, 2018), we ask in this collection how design theories can support social justice advocacy in community-engaged TPC work.

Overview of the Collection

Part 1 of this collection features the intersection of design theory and social justice principles in community-engaged projects. The chapters in this section cover definitions of community engagement, ethics and ethical considerations for community partnership, and methodologies for designing justice-oriented collaborations. To open, Aimée Knight defines community-led design, rather than university-led method, as a collaborative framework that promotes equity, inclusion, and justice. Starting with an asset-driven mentality, this approach leverages the strengths of the community partner and focuses on co-creation in the process of iterating, making, and testing outcomes. Next, Lisa L. Phillips magnifies in her chapter the role of community voices in advancing environmental justice. Her procedural methods apply sensory rhetorics and design activities to redress emergent environmental problems. In another methodological chapter, "Trust, Understand, Act," Erin Brock Carlson teaches us how to carry out rural community projects with visual and place-based research methods that capture multiple public perspectives. The chapter includes guidelines for creating a place-based portfolio. Dovetailing the methodological discussions, Sarah Moon in "Eating to Heal" follows a design thinking approach to redesign a community writing project. The chapter showcases the usefulness of interviews and performance narratives in exploring dietary health issues. Interviewing appears to be a crucial research method in engaged scholarship. Drawing on qualitative interview data with social media content moderators at big tech firms, Steve Holmes and Jared Colton theorize the ethics of constraining user activity via interface design. They suggest a pedagogical framework based on this study to encourage technical communication students to use design as a channel for promoting social justice. Lastly in this section is a proposal on redefining ethical engagement through reflection. Based on their collaborative experience, Sweta Baniya and community leaders offer a retrospective account of the ethical requirements in community-engaged efforts, including the notion of reciprocity in the balancing act of academic and community needs.

Part 2 presents a series of multimodal community-engaged projects that apply design ethics and justice theories to demonstrate the application of these frameworks in situated contexts. First, Felicita Arzu-Carmichael, Therese Pennell, and Josephine Walwema call attention to the current discourse surrounding immigrants, immigration, and specifically, the livelihood of international scholars as a way of opening discussions about community engagement in terms of graduate student socialization. Using a design thinking process, the authors demonstrate how inclusive professionalization can be achieved with a focus on the material conditions of international scholars' lives. In "What We Came Here For," Serenity Hill, Ahlan Filstrup, and Jeanne Beatrix Law showcase their lessons learned from the Atlanta Student Movement Project, a digital, historical civil rights community project, using student reflections collected through interviews. The chapter includes an evaluation of contract-based grading methods as a democratic effort to boost student and community engagement. Next, Jennifer Nish documents the way ME_ Action, an organization that engages in advocacy and activism around myalgic encephalomyelitis (ME) and complex chronic diseases, uses human-centered approaches to design activist activities. Nish foregrounds disability justice principles such as intersectionality and leadership in the effort of diversifying advocacy and activism communities. Combining creativity and advocacy, Laura Gonzales shows how an art installation project was produced and delivered through a partnership between researchers and designers and offers strategies for designing community projects grounded in social justice advocacy. Similar to Gonzales's public project, Tobechukwu Friday, Kristen Tcherneshoff, and Daniel Bögre Udell examine the innovative work of Wikitongues, a nonprofit language revitalization organization, to consider the importance of community-led design efforts in preserving linguistic diversity, safeguarding cultural heritage, and promoting global understanding of languages. Concluding this section is a compelling chapter that features an international project. Liping Yang and Xiaobo Wang present their study of social media influence on a global activist movement. The chapter captures the collective and creative use of social media features in establishing social justice narratives to amplify grassroots voices, mobilize public opinion, and influence policy-making.

Part 3 focuses on pedagogical approaches and teaching recommendations for leveraging multimodal resources and technologies for social justice. Seizing the kairotic moment of online teaching and learning in TPC, Antonio Byrd applies counterstory as a framework for reimagining user persona construction in user-centered design. Byrd describes a technical communication course framed around social justice and creating community partnerships and clarifies the usefulness of counterstory for designing social justice support. In "Designing with Care," Danielle Koepke addresses the efforts and experiences of designing a digital storytelling project about the reproductive justice stories of the Promotores de Salud, Latinx health promoters working for reproductive justice in their communities across Wisconsin. This chapter examines how a cultural rhetoric praxis of care can offer practical actions for caring for vulnerable stories and storytellers that are in line with social justice initiatives within TPC. Next, Carrie Grant illustrates a pedagogical process in which ethical principles were implemented as a core attribute in community design projects. Grant reports on teaching activities and student assignments that connect design thinking to ethical engagement. To advocate for the inclusion of disability justice in TPC pedagogy, Ellen Cecil-Lemkin highlights the outcomes of a service learning project where students collaborated with a service dog nonprofit to create educational social engagement content. The chapter emphasizes inclusive design principles and accessibility in community engagement. Finally, in "Pedagogical Approaches to Normalize Inclusive Design," Jamal-Jared Alexander, Dorcas Anabire, and Rebecca Walton feature a visual design course in which considerations of social justice are intentionally threaded throughout the course to teach students about accessibility, representation, inclusivity, and power relations in design projects.

Together, the 17 chapters in this collection exhibit the many ways in which ethics and design theories can be used to address social justice issues and promote positive change in communities. Through examining various aspects of community-engaged practices, the contributors of these chapters address the opportunities and challenges of design and design methods in community engagement and offer suggestions for promoting social justice through TPC pedagogies.

References

Acharya, K. R. (2022). Promoting social justice through usability in technical communication: An integrative literature review. *Technical Communication, 69*(1), 6–21. https://doi.org/10.55177/tc584938

Agboka, G. Y. (2014). Decolonial methodologies: Social justice perspectives in intercultural technical communication research. *Journal of Technical Writing and Communication, 44*(3), 297–327.

Agboka, G. Y., & Matveeva, N. (Eds.). (2018). *Citizenship and advocacy in technical communication: Scholarly and pedagogical perspectives.* Routledge.

Amidon, T. A., Moore, K. R., & Simmons, M. (2023). Community engaged researchers and designers: How we work and what we need. *Communication Design Quarterly, 11*(2), 5–9.

Armfield, D., Gurak, L., Kays, T. M., & Weinberg, J. (2012). Technical communication education in a digital, visual world. In S. Bartell (Ed.), *Proceedings of the 2012 IEEE international professional communication conference* (pp. 1–5). IEEE. https://doi.org/10.1109/IPCC.2012.6408637

Arola, K. L. (2018). A land-based digital design rhetoric. In J. Alexander, & J. Rhodes (Eds.), *The Routledge handbook of digital writing and rhetoric* (pp. 199–213). Routledge.

Baca, I. (2012). *Service-learning and writing: Paving the way for literacy(ies) through community engagement.* Brill.

Balghare, A. J. (2022). Healthcare communication as a social justice issue: Strategies for technical communicators to intervene. *Present Tense, 9*(3), 1–10. https://www.presenttensejournal.org/volume-9/healthcare-communication-as-a-social-justice-issue-strategies-for-technical-communicators-to-intervene/

Baniya, S., Call, K., Brein, A., & Kumar, R. (2022). COVID-19, international partnerships, and the possibility of equity: Enhancing digital literacy in rural Nepal amid a pandemic. *Reflections: A Journal of Community-Engaged Writing and Rhetoric, 21*(1). https://reflectionsjournal.net/2022/02/covid-19-international-partnerships-and-the-possibility-of-equity-enhancing-digital-literacy-in-rural-nepal-amid-a-pandemic/

Bay, J. (2022). Fostering diversity, equity, and inclusion in the technical and professional communication service course. *IEEE Transactions on Professional Communication, 65*(1), 213–225. https://ieeexplore.ieee.org/document/9708711

Bay, J., Johnson-Sheehan, R., & Cook, D. (2018). Design thinking via experiential learning: Thinking like an entrepreneur in technical communication courses. *Programmatic Perspectives, 10*(1), 172–200.

Benjamin, R. (2019). *Race after technology: Abolitionist tools for the new Jim code.* Polity.

Bennett, K. C., & Hannah, M. A. (2021). Generative fusions: Integrating technical and professional communication, disability studies, and legal studies in the work of disability inclusion and access. *IEEE Transactions on Professional Communication, 64*(3), 235–249. https://ieeexplore.ieee.org/document/9509583

Blythe, S., Grabill, J., & Riley, K. (2008). Action research and wicked environmental problems: Exploring appropriate roles for researchers in professional communication. *Journal of Business and Technical Communication, 22*(3), 272–298. https://doi.org/10.1177/1050651908315973

Breaux, C. (2017). Why making? *Computers and Composition, 44*, 27–35.

Cagle, L. E. (2017). Becoming "forces of change": Making a case for engaged rhetoric of science, technology, engineering, and medicine. *POROI, 12*(2), 1–13. https://doi.org/10.13008/2151-2957.1260

Carlson, E. B. (2020). Embracing a metic lens for community-based participatory research in technical communication. *Technical Communication Quarterly*, *29*(4), 392–410. https://doi.org/10.1080/10572252.2020.1789745

Carnegie Foundation for the Advancement of Teaching. (2006). The elective classification for community engagement: What is community engagement? Carnegie Classification of Institutions of Higher Education. https://carnegieclassifications.acenet.edu/elective-classifications/community-engagement/

Colton, J. S., & Holmes, S. (2018). A social justice theory of active equity for technical communication. *Journal of Technical Writing and Communication*, *48*(1), 4–30. https://doi.org/10.1177/0047281616647803

Cooke, L., Dusenberry, L., & Robinson, J. (2020). Gaming design thinking: Wicked problems, sufficient solutions, and the possibility space of games. *Technical Communication Quarterly*, *29*(4), 327–340.

Costanza-Chock, S. (2020). *Design justice: Community-led practices to build the worlds we need*. MIT Press. https://mitpress.mit.edu/9780262043458/design-justice/

Cox, M. B. (2019). Working closets: Mapping queer professional discourses and why professional communication studies need queer rhetorics. *Journal of Business and Technical Communication*, *33*(1), 1–25. https://doi.org/10.1177/1050651918798691

Dark, T., & Baker, W. D. (2015). Entering the conversations, practices and opportunities of multimodality texts. *Teaching/Writing: The. Journal of Writing Teacher Education*, *4*(1), 65–93. https://scholarworks.wmich.edu/wte/vol4/iss1/4

Deans, T., Roswell, B., & Wurr, A. J. (2010). *Writing and community engagement: A critical sourcebook*. Bedford/St. Martin's.

Dubinsky, J., & Carpenter, J. H. (2004). Special issue: Civic engagement and technical communication. *Technical Communication Quarterly*, *13*(3), 1–128.

Duin, A. H., & Pedersen, I. (2023). *Augmentation technologies and artificial intelligence in technical communication: Designing ethical futures*. Routledge.

Edenfield, A. C. (2017). Power and communication in worker cooperatives: An overview. *Journal of Technical Writing and Communication*, *47*(3), 260–279. https://doi.org/10.1177/0047281616641921

Edenfield, A., & Ledbetter, L. (2019). Tactical technical communication in communities: Legitimizing community-created user-generated instructions. In *Proceedings of the 37th ACM international conference on design of communication* (pp. 1–9). https://doi.org/10.1145/3328020.3353927

Eubanks, V. (2012). *Digital dead end: Fighting for social justice in the information age*. MIT Press.

Eubanks, V. (2018). *Automating inequality: How high-tech tools profile, police, and punish the poor*. St. Martin's Press.

Eyman, D., & Ball, C. (2014). Composing for digital publication: Rhetoric, design, code. *Composition Studies*, *42*(1), 114–117.

Grabill, J. (2007). *Writing community change: Designing technologies for citizen action*. Hampton Press.

Garza, S. (2013). *Adding to the conversation on service-learning in composition: Taking a closer look*. Fountainhead Press.

Haas, A. (2012). Race, rhetoric, and technology: A case study of decolonial technical communication theory, methodology, and pedagogy. *Journal of Business and Technical Communication*, *26*(3), 277–310. https://doi.org/10.1177/1050651912439539

Haas, C. (1996). *Writing technology: Studies in the materiality of literacy*. Lawrence Erlbaum.

Holmes, A. (2016). *Public pedagogy in composition studies*. National Council of Teachers of English.

Itchuaqiyaq, C. U., & Matheson, B. (2021). Decolonizing decoloniality: Considering the (mis) use of decolonial frameworks in TPC scholarship. *Communication Design Quarterly*, 9(1), 20–31. https://doi.org/10.1145/3437000.3437002

Jiang, J. (2024a). Composing to enact affective agency: Engaging multimodal anti-racist pedagogy in the first-year writing classroom. *College Composition and Communication*, 75(3), 534–557.

Jiang, J. (2024b). "Emotions are what will draw people in": A study of critical affective literacy through digital storytelling. *Journal of Adolescent & Adult Literacy*, 67(4), 253–263. https://doi.org/10.1002/jaal.1322

Jiang, J., & Tham, J. (2019). Multimodal design and social advocacy: Charting future directions for design as an interdisciplinary engagement. *Digital Rhetoric Collaborative*. http://www.digitalrhetoriccollaborative.org/2019/02/05/multimodal-design-social-advocacy/

Jiang, J., & Tham, J. (2023). Rethinking multimodal community-engaged pedagogy through posthumanist theory. *Teaching in Higher Education*. 1–17. https://doi.org/10.1080/13562517.2023.2253758

Jiang, J., & Tham, J. (2024). Race, affect, and marginalized communities: Navigating racialized emotions in community-engaged pedagogy. *Critical Studies in Education*, 1–19. https://doi.org/10.1080/17508487.2024.2343284

Jones, N. N. (2016). The technical communicator as advocate: Integrating a social justice approach in technical communication. *Journal of Technical Writing and Communication*, 46(3), 342–361. https://doi.org/10.1177/0047281616639472

Kalantzis, M., & Cope, B. (2012). *Literacies*. Cambridge University Press.

Kimbell, L. (2011). Rethinking design thinking: Part 1. *Design and Culture*, 3(3), 285–306. https://doi.org/10.2752/175470811X13071166525216

Kimbell, L. (2012). Rethinking design thinking: Part 2. *Design and Culture*, 4(2), 129–148. https://doi.org/10.2752/175470812X13281948975413

Knight, A. (2022) *Community is the way: Engaged writing and designing for transformative change*. The WAC Clearinghouse; University Press of Colorado. https://wac.colostate.edu/books/practice/community/

Kress, G. (2010). *Multimodality: A social semiotic approach to contemporary communication*. Routledge.

Leydens, J. A. (2012). What does professional communication research have to do with social justice? Intersections and sources of resistance. In *Proceedings of the 2012 IEEE international professional communication conference* (pp. 1–13). https://ieeexplore.ieee.org/document/6408592

Mara, A. (2006). Using charrettes to perform civic engagement in technical communication classrooms and workplaces. *Technical Communication Quarterly*, 15(2), 215–236.

Matthews, C., & Zimmerman, B. (1999). Integrating service learning and technical communication: Benefits and challenges. *Technical Communication Quarterly*, 8(4), 383–404. https://doi.org/10.1080/10572259909364676

Moore, K. R. (2017). The technical communicator as participant, facilitator, and designer in public engagement projects. *Technical Communication*, 64(3), 237–253.

Palmeri, J. (2012). *Remixing composition: A history of multimodal writing pedagogy*. SIU Press.

Patterson, L. (2015). A quantitative discourse analysis of first year engineering student reflections: A pilot study of a service learning communication assignment. In N. Werner (Ed.), *Proceedings of IEEE international profes- sional communication conference* (pp. 1–2). IEEE. https://doi.org/10.1109/ IPCC.2015.7235794

Pellegrini, M. (2021). Composing like an entrepreneur: The pedagogical implica- tions of design thinking in the workplace. *Journal of Technical Writing and Com- munication, 35*(2), 185–218. https://doi.org/10.1177/00472816211031554

Rivera, N. K., & Gonzales, L. (2021). Community engagement in TPC programs during times of crises: Embracing Chicana and Latina feminist practices. *Pro- grammatic Perspectives, 12*(2), 39–65.

Sackey, D. (2018). An environmental justice paradigm for technical communi- cation. In A. Haas, & M. Eble (Eds.), *Key theoretical frameworks: Teaching technical communication in the twenty-first century* (pp. 138–160). Utah State University Press.

Sapp, D. A., & Crabtree, R. D. (2002). A laboratory in citizenship: Service learn- ing in the technical communication classroom. *Technical Communication Quarterly, 11*(4), 411–432. https://doi.org/10.1207/s15427625tcq1104_3

Selber, S. (2004). *Multiliteracies for a digital age*. Southern Illinois UP.

Shah, R. (2020). *Rewriting partnerships: Community perspectives on community- based learning*. Utah State University Press.

Sheridan, D. M., Ridolfo, J., & Michel, A. J. (2012). *The available means of per- suasion: Mapping a theory and pedagogy of multimodal public rhetoric*. Parlor Press.

Shivers-McNair, A. (2017). Localizing communities, goals, communication, and inclusion: A collaborative approach. *Technical Communication, 64*(2), 97–112.

Simmons, W. M. (2007). *Participation and power: A rhetoric for civic discourse in environmental policy*. SUNY Press.

Simmons, W. M., & Amidon, T. R. (2019). Negotiating research stance: An ecol- ogy of tensions in the design and practice of community-engaged research. In *Proceedings of the 37th ACM international conference on the design of com- munication* (pp. 1–11).

Spilka, R. (2009). *Digital literacy for technical communication: 21st century the- ory and practice*. Routledge.

Takayoshi, P., & Selfe, C. L. (2007). Thinking about multimodality. In C. L. Selfe (Ed.), *Multimodal composition: Resources for teachers* (pp. 1–28). Hampton Press.

Tham, J. (2021). *Design thinking in technical communication: Solving problems through making and collaboration*. Routledge.

Tham, J., & Grace, R. (2024). Design thinking in localized service-learning: In- novating solutions through empathy, research, and advocacy. In A. Lancaster, & C. King (Eds.), *Amplifying voices in UX: Balancing design and user needs in technical communication* (pp. 21–47). SUNY Press.

Tham, J., & Jiang, J. (2022). Examining multimodal community-engaged pro- jects for technical and professional communication: Motivation, design, tech- nology, and impact. *Journal of Technical Writing and Communication, 53*(2), 128–159. https://doi.org/10.1177/00472816221115141

Tham, J., & Jiang, J. (2024). Understanding writing instructors' feelings toward the affordances of multimodal social advocacy projects: Implications for service-learning pedagogies. *College Composition and Communication, 76*(1), 4–34.

Turnley, M. (2007). Integrating critical approaches to technology and service- learning projects. *Technical Communication Quarterly, 16*(1), 103–123. https://doi.org/10.1080/10572250709336579

van Leeuwen, T. (2005). Multimodality, genre, and design. In R. H. Jones & S. Norris (Eds.), *Discourse in action: Introducing mediated discourse analysis* (pp. 73–93). Routledge. https://doi.org/10.4324/9780203018767

Walton, R., & Agboka, G. Y. (Eds.). (2021). *Equipping technical communicators for social justice work: Theories, methodologies, and pedagogies.* Utah State University Press.

Walton, R., & Hopton, S. B. (2018). "All Vietnamese men are brothers": Rhetorical strategies and community engagement practices used to support victims of Agent Orange. *Technical Communication, 65*(3), 309–325.

Walton, R., Moore, R., & Jones, N. N. (2019). *Technical communication after the social justice turn: Building coalitions for action.* Routledge.

Wible, S. (2020). Using design thinking to teach creative problem solving in writing courses. *College Composition and Communication, 71*(3), 399–425.

Wysocki, A. F., & Lynch, D. A. (2018). *Compose, design, advocate: A rhetoric for multimodal communication* (3rd ed.). Pearson.

Zdenek, S. (2020). Transforming access and inclusion in composition studies and technical communication. *College English, 82*(5), 536–544.

PART I
Theories and Ethics in Designing for Social Justice

1

COMMUNITY-LED DESIGN

Building Frameworks for Equity and Justice

Aimée Knight

Introduction

Community-led design is an invitation to those involved in developing more inclusive frameworks for equity and justice. A community-led approach shifts power dynamics and places the community at the center of the design research process. Here, I advocate for this grassroots approach to designing community-university partnerships—an approach I am exploring while directing media projects in writing and design while teaching two writing courses: Nonprofit Communication and Media & Engagement. The Beautiful Social Research Collaborative serves as the coordinating infrastructure for this work. Through a process of co-creation, nonprofit and partner organizations receive support on an issue they identify while students gain valuable experience putting theory into practice. Co-creation here involves the collaborative generation of new knowledge between university and community partners. Co-creation emphasizes shared authority, reciprocal learning, and the acknowledgment of community expertise, fostering a shared approach where power and decisions are shared in all aspects of the relationship.

At the Beautiful Social Research Collaborative, we have engaged in over 130 projects in professional and technical writing, social media strategy and content creation, web design, podcast production, and web-based video free of charge with communities in the Greater Philadelphia Area and countries around the world including Haiti, Kenya, Sierra Leone, and The Gambia. Relevant to educators, civic organizations, nonprofits, and activists, the chapter serves as a guide for implementing a community-led design

DOI: 10.4324/9781003469995-3

approach that demonstrates "collective effort and cooperation, not individualism and competition" (Russell & McKnight, 2022, p. xiv).

Empowering Communities through Community-Led Design

In community-university partnerships, there is a growing emphasis on achieving a positive impact for our civic partners. However, realizing this impact remains a complex challenge, partly due to the historical (and often unintentional) missteps of universities in these relationships (Yates & Accardi, 2019, p. 6). Historically, community-university partnerships have tended to prioritize academic concerns—while often minimizing citizen interests and expertise. Writing studies and community engagement scholars have challenged the prevailing nature of community-university partnerships and how they often prioritize the university's interests over those of the community (Bortolin, 2011, p. 55). Researchers observing this trend note that universities tend to focus on validating the discipline itself and frequently exercise more control in partnerships—privileging student outcomes, faculty perceptions, and academic timelines in their research. Furthermore, scholars have pointed out the conspicuous absence of the community in the existing literature (Ball & Goodburn, 2000, p. 82). Scholars have questioned whether—or to what extent—community members even derive tangible benefits from these partnerships and whether universities and communities can genuinely collaborate for mutual benefit (Yates & Accardi, 2019, p. 44).

Addressing these concerns requires us to redesign both structures and mindsets regarding community-university partnerships. The adoption of methods such as asset-based community development (Cruz & Giles, 2000; Saltmarsh & Hartley, 2012; Shah et al., 2018) offers a promising pathway forward. This approach, embraced by writing studies and the community engagement field, emphasizes building on the strengths and assets of communities (Kretzmann & Mcknight, 1993). Unlike deficit-based views that stigmatize communities as fundamentally lacking or problematic, asset-based community development highlights the positive capacities of communities, such as creativity, local wisdom, and resilience. A hallmark of the asset-based approach is the focus on "what a community has that can be further developed and utilized by the community" (Cruz & Giles, 2000, p. 31).

The adoption of an asset-based approach also challenges scholarly bias. Rather than framing research as a response to problems, it prioritizes utilizing local assets to tackle inequalities (Harrison et al., 2019). This shift encourages the development of mutual respect and trust between universities and communities. Kreztmann and McKnight (1996) recommend a focus on relationship building and emphasize the need to "constantly build and re-build the relationships between and among local residents, associations, and institutions" (p. 27). By placing the community's gains first, this approach reframes research as a process that builds and strengthens communities.

A complementary approach lies in community-led design, an effective method for addressing inequalities and fostering mutual respect. Community-led design is an inclusive and participatory approach to designing projects and initiatives that directly involves the members and stakeholders who are most affected by the outcomes. The community-led design movement emerged in the 1960s as a participatory design initiative in urban design, planning, and architecture (Sanoff, 2011). The movement was a response to an increased sense of social responsibility and "a growing sense that people should have the right to participate directly in shaping the environments they live in" (Alexiou et al., 2015, p. 31). The essence of community-led design lies in its commitment to the "by us, for us" approach, where "designs are created by community members for themselves, rather than having designs created for them by others" (Introduction to Community-Led Co-Design, n.d.). This approach aims to empower communities to have an active role in shaping the processes and impact of the research, rather than being passive recipients. Table 1.1 provides an overview of the different levels of engagement in partnerships ranging from community-owned to university owned.

Community-led design involves the collaborative development of the design process, ensuring that community members directly impacted by the research are actively guiding the process. Traditional university-led models can perpetuate colonial notions of ownership, control, and the pursuit of institutional status (such as grants, publications, tenure, and promotion). Rooted in power dynamics, a traditional approach risks marginalizing voices and reinforcing paternalistic views of the community. In contrast, a community-led framework focuses on joint ownership and co-creation, allowing for more equitable partnerships where communities are active agents rather than passive recipients. This shift in dynamics contributes to citizen empowerment by valuing local assets and wisdom, growing relationships built on trust, sharing power, and ensuring that the community's gains are prioritized over institutional gains. Table 1.2 provides a further comparison highlighting key characteristics, guiding principles, and outcomes.

Community-Led Approach Self-Assessment

This assessment tool provides some clarity about the values and priorities that shape your current approach. The scoring system will help to interpret the results, shedding light on whether the initiative is grounded in community gains or driven by university-led perspectives.

TABLE 1.1 Spectrum of community–university engagement in partnerships.

	Community owned	Community led	University led	University owned
Visions and goals	Local visions for change are defined and implemented by the community.	Local visions for change are created in partnership with community members and organizations (such as nonprofits or universities).	Research and goals are primarily created and driven by university members in collaboration with the community.	Research and goals are primarily created by university members who control resources, parameters, and decisions.
Approach	Citizen-driven change; community members are in control of all resources, parameters, and decisions.	Community members and university members share resources, collaborate on decision-making, and set project parameters.	University members share resources and collaborate with the community on decision-making.	University members may seek community input, but decision-making authority remains with the university.
Collaboration	University members could serve as supportive collaborators and resources for community initiatives.	Emphasizes co-creation, joint ownership, and equitable influence in the partnership.	Community input is considered but may only partially define or guide the direction and scope of the project.	Community involvement is more consultative than collaborative.

TABLE 1.2 Community-led versus university-led approaches.

Aspect	Community-led approach	University-led approach
Context	The program or project prioritizes visions, strengths, and assets of the community partners.	The program or project centers on university-led research or problems in the community (deficit, problem-based).
Focus	The focus is on building community through a collaborative process.	The community is seen as a site or laboratory for research or the project addresses community problems based on academic concerns.
Priority	Community gains are prioritized.	University gains and student-learning outcomes are prioritized.
Nature of collaboration	The collaboration involves co-creation and joint ownership of work.	The university has more agency and dictates terms and timelines.
Relationship dynamics	The relationship is built on a trusting and mutually enriching partnership.	Power dynamics are at play or paternalistic views of the community are present.
Timeframe	The timeframe involves a long-ranging, ongoing investment.	Guided by semester schedules, focuses on short-term projects.
Resource utilization	Utilizes local assets for tackling inequalities.	Controls money, institutional resources.
Evaluation criteria	Evaluation criteria focus on community impact and generativity.	Evaluation criteria privilege faculty perceptions, student learning outcomes, and pursuit of grant funding.
Communication style	Communication is open, collaborative, and trusting.	Discourses are tied to power; communication is driven by the university.
Community engagement approach	The approach is based on an asset-based community development (ABCD) approach, focused on the goals of the community.	The approach is problem or research based, or focused on validating a scholarly discipline.
Guiding questions	Guiding questions include: "Did we engage in a process that builds community?"	Guiding questions include: "What are the community needs or problems we can address?"
Long-term goals	Long-term goals include sustainable community development and empowerment.	Long-term goals revolve around the pursuit of status, promotion, and publication.

Instructions

For statements 1–12, choose the response that best fits your approach. Then add up the points and use the key to interpret your results.

1 Context

- The program or project prioritizes the visions, strengths, and assets of community partners (3 points)
- The program or project moderately prioritizes community visions and strengths (2 points)
- The program or project centers on university-led research (1 point)

2 Focus

- The focus is on building community through collaborative processes (3 points)
- The focus moderately highlights community building (2 points)
- The community is perceived as a laboratory for research, or the project addresses community needs based on academic concerns (1 point)

3 Prioritization

- Community gains are prioritized first (3 points)
- Community and university goals are balanced (2 points)
- University gains and student learning outcomes are put first (1 point)

4 Nature of collaboration

- The collaboration involves co-creation and joint ownership of work (3 points)
- The university involves the community in decision-making (2 points)
- The university has more agency and dictates terms and timelines (1 point)

5 Relationship dynamics

- The relationship is built on a trusting and mutually enriching partnership (3 points)
- The partnership is moderately trusting and mutually enriching (2 points)
- There are some power dynamics at play (such as the university is viewed as the expert) or there are paternalistic views of the community present (1 point)

6 Timeframe

- The timeframe values a long-ranging, ongoing investment; slow paced (3 points)

- The timeframe balances short-term and long-term goals (2 points)
- The timeframe is by semester schedules or short-term projects; fast paced (1 point)

7 Resource utilization

- Utilizes local assets for tackling inequalities (3 points)
- Moderately considers local assets (2 points)
- University controls money or institutional resources (1 point)

8 Evaluation criteria

- The evaluation criteria focus on community impact (3 points)
- The evaluation criteria balances community and institutional goals (2 points)
- Evaluation criteria privilege faculty perceptions, student learning outcomes, or the pursuit of grants, awards, and publications (1 point)

9 Communication style

- Open, collaborative, trusting, community led (3 points)
- University-led but values community expertise (2 points)
- Discourses tied to power; communication is driven by the university (1 point)

10 Community engagement approach

- The approach is based on an equity-based framework and/or focused on the visions of the community (3 points)
- Incorporates some community engagement principles (2 points)
- The approach is problem-based; validates an academic discipline or criteria (1 point)

11 Guiding questions

- "Did we engage in a process that builds community?" (3 points)
- Considers community impact in the project's guiding question (2 points)
- "What are the community needs or problems we can address?" (1 point)

12 Long-term goals

- Long-term goals include sustainable community development and empowerment (3 points)
- Balances sustainable community development with institutional goals (2 points)
- Long-term goals revolve around the pursuit of status, promotion, and publication (1 point)

Key

28–36 points: The approach is strongly community-led, prioritizing collaboration, community strengths, and long-term sustainability. Aligning with the community's vision ensures the project is relevant, empowering, and ultimately valuable for the community.

20–27 points: The approach moderately embraces community-led principles but will benefit from further integration of community priorities. Recognizing and integrating the community's vision and assets can strengthen the project's impact, fostering a deeper connection with the community.

12–19 points: The approach leans toward a more traditional university-centric model. Explore community-led strategies to enhance inclusivity and collaboration. Better integrating the community visions creates mutual ownership and a more sustainable impact.

A Community-Led Design Framework

Community-led strategies are tools to help dismantle power imbalances inherent in traditional university-centric models. While many community-engaged projects will tout a well-versed "mutual and reciprocal relationship" between institutions and communities, a community-led approach ensures that the partnership will ultimately yield a positive community impact. In my work directing writing and media partnerships, I have repeatedly encountered the need for more generative and pragmatic approaches that challenge the status quo and support more authentic possibilities for co-creation with community partners. The framework that follows, illustrated in Figure 1.1, is a flexible and practical guide for implementing community-led design that features equity-based and co-creation tactics.

Equity-Based Approaches

Equity-based approaches lay the foundation for an inclusive and equitable community-led design process. These approaches prioritize empathy and understanding, actively listening to a diversity of voices and the redistribution of power within the design research framework. Although the steps are numbered below, the emphasis is on a nonhierarchical and integrative framework.

(1) Identify the Community

Understand the community we are working with and identify diverse stakeholders, including community members, organizations, educators, activists, and practitioners invested in the project or partnership.

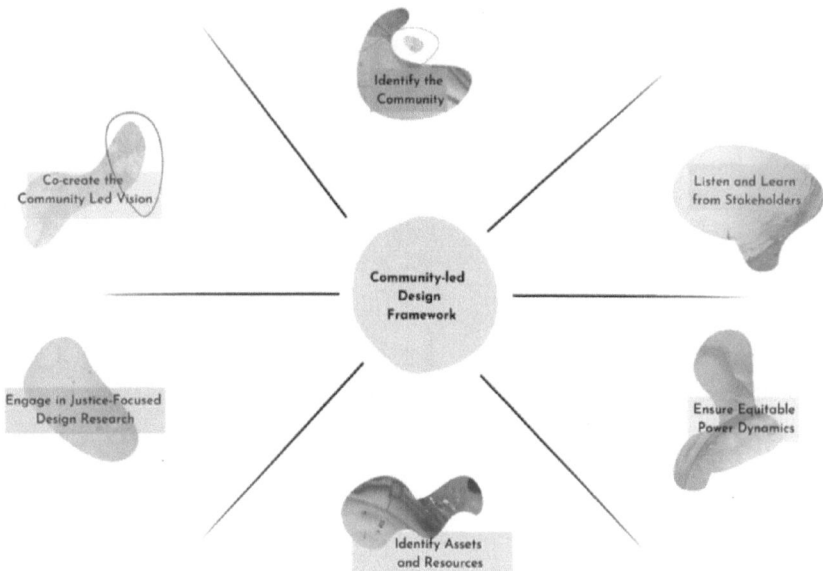

FIGURE 1.1   Community-led design framework.

Identifying the community is a foundational step (and is often taken for granted). To move from a vague to a clear understanding of the community, we can engage in comprehensive research and dialogue. By identifying and creating a working understanding of the community, we have an appreciation for its unique identity and can tailor our approach accordingly. This step ensures the design process is inclusive and tailored to its specific needs.

Guiding question: How can we understand the diverse elements that make up the community? Some hands-on ways to enhance our understanding of the community include:

- **Community immersion:** Spend time in the community by actively participating in a local event, volunteering, or visiting the site or neighborhood to gain a deeper understanding of the community and its values.
- **Goal mapping:** Visually map out shared goals and aspirations of the project or partnership. Goal mapping aligns resources and efforts around common objectives, providing a roadmap for the design process. It serves as a guide for assessing impact, ensuring that the community's vision remains the focus throughout the project.

FIGURE 1.2 AnaOno goal map.

Figure 1.2 is a goal map example from the AnaOno student group. This group aimed to develop a versatile media strategy focused on the breast cancer community, emphasizing flexibility and responsiveness. Through collaboration, they co-created resources and worked to strengthen brand visibility and engagement while aligning closely with AnaOno's mission to cater to diverse chest types and empower those impacted by breast cancer.

According to the student team: "The goal mapping activity is a useful tool in order to visually represent the shared goals of our partnership with AnaOno. As we recognize the initial and revised goals over the course of the semester, we are able to strengthen our next steps and ensure result-oriented deliverables."

A goal-mapping design activity is shared at the end of this chapter.

(2) Listen and Learn from Stakeholders

Active listening is a practice that empowers and amplifies community voices in the design process. It brings present-moment awareness and deep listening skills to our encounters.

It is essential to bring present-moment awareness and deep listening skills to our encounters with community partners. Active listening is a practice that empowers and amplifies community voices in the design process. Robin Wall Kimmerer emphasizes the significance of listening when reflecting on the Onondaga Nation's Thanksgiving Address or "The Worlds That Come Before All Else." This practice involves reciprocating spoken words with attention:

> The listeners reciprocate the gift of the speaker's words with their attention, and by putting their minds into the place where gathered minds meet. You could be passive and just let the words and time flow by, but each call asks for the response "Now are minds are one." You have to concentrate; you have to give yourself up to the listening. It takes effort, especially in a time when we are accustomed to sound bites and immediate gratification.
>
> *(Kimmerer, 2022, p. 106)*

Guiding question: How can we bring present-moment awareness and deep listening skills to our encounters with communities? Consider the following methods:

- **Reflective listening:** Practice non-judgmental listening and reflecting back to enhance communication. Reflective listening includes "responding to the other person by reflecting the thoughts and feelings you heard in his or her words, tone of voice, body posture, and gestures" (Katz & McNulty, 1994, p. 1). Inclusion expert Pearpoint suggests that we also need to "build in time to process what it was like to listen and to be listened to. Many of us in North America find listening to one another difficult—we want to jump in, interrupt or add pearls of wisdom" (Forest & Pearpoint, 1995).
- **Social media/digital engagement.** Listen (with permission) to the community's online presence by joining social media groups, online forums, or platforms where individuals share their perspectives.

(3) Ensure Equitable Power Dynamics

Mindfully address power imbalances and work to create an equitable decision-making process. This step involves holding open conversations about power dynamics and actively working to dismantle them. This step establishes a more just and balanced foundation for decision-making, ensuring that all voices hold weight.

Part of our work in the design process is to challenge and dismantle dominant cultural mindsets. In her book, *Decolonizing Design: A Cultural Justice Guidebook,* Dori Dunstall contends that the dominance of white dominant culture in many areas is a result of European colonization (Tunstall, 2023, p. 78). When we work to ensure equitable power dynamics in our project design, we actively promote equity and the dismantling of white supremacy culture. The Design Justice Network argues that "the people who are most adversely affected by design decisions—about visual culture, new technologies, the planning of our communities, or the structure of our political and economic systems—tend to have the least influence on those decisions and how they are made" (Design Justice Principles Overview). This approach works to address historical inequities, empower marginalized communities, and respect multiple ways of knowing and being.

Guiding question: In what ways can power dynamics be addressed and equitable decision-making processes established? To strive for equitable power dynamics in a community-led design framework, consider the following methods:

- **Uncover implicit associations:** Take and reflect on The Implicit Association Tests (IAT) from Project Implicit which can bring implicit biases to light and foster a more inclusive and just design process.
- **Social change ecosystem framework:** The Social Change Ecosystem Framework helps individuals identify which roles they are best suited for in social change initiatives, including visionary, builder, disruptor, experimenter, storyteller, and healer. Deepa Iyer argues that "organizations and networks can use the framework to deepen collaboration, coordination, and connections, to understand how their various members and components are working together to advance common goals" (Iyer, 2022, p. 19).
- **Equity-based training:** Reflect on "White Supremacy Culture Characteristics" with students, faculty, and other project stakeholders. Tema Okun (2021) contends that white supremacy culture trains us all to internalize attitudes and behaviors that do not serve any of us (White Supremacy Culture Characteristics). The list of white supremacy characteristics includes perfectionism, a sense of urgency, defensiveness, valuing quantity over quality, worship of the written word, belief in only one right way, paternalism, either/or thinking, power hoarding, fear of open conflict, individualism, and claiming a right to comfort.

Co-creation Approaches

When we focus on co-creation approaches, we recognize that the process of the community-university collaboration can be more important than the ultimate impact of the project. Featuring a community-led practice of iterating, making, testing, and community-led design, the focus of the work is on how we can collaboratively develop a shared process that reflects the community's goals and aspirations.

(4) Identify (Existing) Assets and Resources

Explore the existing assets and resources within the community. This includes expertise, networks, facilities, skills, and knowledge that can contribute to the success of the project.

An asset, according to the "Mapping Community Assets Workbook," is "an item of value owned; a quality, condition, or entity that serves as an advantage, support, resource, or source of strength" (Dorfman, 1998). This approach celebrates what assets the community already possesses rather than focusing on what it might lack and can be framed as moving from a "what's wrong to what's strong" position (Russell & McKnight, 2022, p. 31). Recognizing existing assets and resources within the community is an act of both respect and empowerment.

Guiding question: What existing assets and resources within the community can be identified and leveraged to support the success of the community-led design initiative? Consider the following methods:

- **Community asset mapping:** Organize a meeting involving community members, organizations, or practitioners to map and document existing local assets and resources. By strategically leveraging these resources, we ensure that the design process is grounded in the community's capabilities, making it more sustainable and less reliant on external support.
- **Community asset survey:** Conduct a survey or questionnaire to identify and document existing assets and resources in key categories such as: environmental, built, social, economic, and public. A survey serves as a tool to uncover valuable resources, skills, and networks that may otherwise take time to become apparent.

(5) Engage in Justice-focused Design Research

Employ a justice-focused design research model such as co-design, design justice, or an equity-based approach to challenge traditional methods, (such as design thinking) and promote more inclusive practices.

Embracing justice-focused design research is an act of resistance against traditional, often white-dominant approaches.

Guiding question: How can we incorporate design models, such as co-design, design justice, and equity-based methods, to challenge the status quo and promote more inclusive practices? Some approaches include:

- **Co-creation workshops:** Organize workshops where community members, designers, and stakeholders come together to brainstorm, ideate, and create the vision for the design approach. Design models such as co-design, design justice, community-led design, and equity-based approaches help to challenge the status quo and encourage diverse perspectives.

- **Prototyping and beta testing:** An iterative design process creates infrastructure for meaningful collaboration. Actively involve stakeholders in the creation, development, and testing of prototypes or beta versions. For ideation and prototype approaches see the "Equity-Centered Community-Design Field Guide" by the Creative Reaction Lab.
- **Frequent check-ins:** Schedule frequent virtual check-ins via accessible platforms such as Zoom to provide timely updates, gather feedback, and collaboratively adapt and evaluate the design. Regular feedback sessions help to maintain transparency and offer stakeholders ongoing opportunities for input. Frequent conversations grow trust and transparency. Rotate shorter virtual check-ins with longer in-person meetings.

(6) Co-create the Community-Led Vision

A successful project depends on its responsiveness to the community's vision. Remain aligned with the community's goals and continuously evaluate the community-led design approach.

Regularly seek feedback from all stakeholders to keep in sync with the community's aspirations for the project.

Guiding question: How can we ensure that the design process is rooted in shared values and an authentic commitment to the community's vision? To further develop the community-led design process, consider the following methods:

- **Shared principles:** Generate a set of guiding principles for future work based on common core values in the group. Look to the Design Justice Network https://designjustice.org/principles-overview and the Allied Media Network https://alliedmedia.org/network-principles for examples.
- **Evaluation session:** Administer surveys or host a feedback session to gather insights, preferences, and concerns from community members. Inquire not only about the project impact but also about the co-creation process itself. Create a platform where all voices are heard and actively integrated into the conversation.
- **Community party:** Organize an event where the community and university come together to showcase their projects and share stories. Have a party, not a meeting. For example, at the Beautiful Social Research Collaborative, we opt for a community party at the end of each semester, where university and community stakeholders physically come together to celebrate our projects, strengthen our ties, and share delicious food.

A party builds trust and fosters a sense of ownership among all participants. In a later meeting, we might undergo a more formal evaluation process, discuss the project impact, and reflect on areas for improvement or future directions. By consistently nurturing relationships and fostering open communication with community partners, we can ensure that our projects not only succeed but become deeply relevant to the lived experiences and goals of the communities they aim to benefit.

Transforming Communities through Design

What does the community-led design framework accomplish that traditional methods cannot? Community-led design provides practitioners with a roadmap for implementation that centers community-driven change. A community-led design framework supports communities in becoming co-authors of their own visions. It recognizes that design research is not a one-size-fits-all endeavor; it's a collaborative, creative process that actively involves those it serves. In contrast, some traditional design research methods can perpetuate power imbalances, discrimination, and cultural insensitivity. The community-led design framework represents a commitment toward more equitable, inclusive, and justice-oriented design processes. It recognizes that every successful design lies within the heart of the community it serves.

At its core, community-led design explores collaborative and inclusive practices. By actively engaging with the community, it goes beyond the surface level, delving into the systemic challenges the community faces. It promotes deep listening to stakeholders and weaving their unique expertise into the project design. Equity-based approaches leverage existing assets, foster collaboration, and ensure equitable power dynamics. Co-creation approaches encourage design research models like co-design, design justice, and equity-based approaches, which champion inclusive and diverse perspectives. The community-led design framework embraces an iterative process of evaluation and adaptation while soliciting consistent feedback. It is not a static model but a living, responsive process that positions community stakeholders as co-creators of their critical futures.

Design Activity: Goal-Mapping Your Partner's Vision

As seen in Figure 1.2, a goal map visually represents the shared goals and aspirations of the project or partnership. This map should serve as a roadmap for the design process, aligning resources and efforts around common objectives. The activity emphasizes the importance of integrating the community's voice into the project and keeps us on track for the evaluation process. It is important to revisit these goals early and also at mid-semester to stay on track and ensure project success.

1 Understanding the importance of goal mapping: In project groups, begin by reflecting on the importance of goal mapping in the community-led design process. Consider how goal mapping can help keep your project on track, ensure alignment with the community's vision, and serve as a tool for project evaluation.
2 Revisiting project goals and aspirations: Take some time to revisit the stated goals, visions, and aspirations of your project or partnership—i.e., those of your community partner. Reflect on previous conversations with your partner and any feedback you have received throughout the semester.
3 Creating your goal map: Use your reflection to sketch a visual goal map that represents the shared goals and aspirations of your entire project from start to finish. You can choose from various visualization techniques, such as mind maps, timelines, flowcharts, diagrams, or something more homegrown, to illustrate the goal map.
4 Integrating the community's voice: Is your project adhering to the community's vision or is it beginning to go off course? It is essential to incorporate the community's voice and vision into your goal map, based on insights derived from previous interactions and discussions. Consider all of the needs, preferences, and aspirations of the community as you create your map.
5 Presenting your goal map: Once your goal map is complete, you will have the opportunity to present it to your peers. During the discussion, reflect on how the goal map will guide the remaining stages of your project.
6 Report: You can include this map in a larger community partner report to illustrate the design of your project.

References

Alexiou, K., Alevizou, K., Zamenopoulos, T., deSousa, S., & Dredge, L. (2015). Learning from the use of media in community-led design projects. *Cultural Science Journal, 8*(1), 30–40. https://doi.org/10.5334/csci.71

Ball, K., & Goodburn, A. M. (2000). Composition studies and service learning: Appealing to communities? *Composition Studies, 28*(1), 79–94.

Bortolin, K. (2011). Serving ourselves: How the discourse on community engagement privileges the university over the community. *Michigan Journal of Community Service Learning, 18*(1), 49–58.

Costanza-Chock, S. (2020). *Design justice: Community-led practices to build the worlds we need*. MIT Press.

Creative Reaction Lab. (2018). *Field guide: Equity-centered community design*. Creative Reaction Lab. https://www.creativereactionlab.com/store/p/field-guide-equity-centered-community-design

Cruz, N., & Giles, D. (2000). Where's the community in service-learning research? *Michigan Journal of Community Service Learning, 1,* 28–34.

Dorfman, D. (1998). *Mapping community assets workbook.* Northwest Regional Educational Laboratory. https://resources.depaul.edu/abcd-institute/resources/Documents/DorfmanMappingCommunityAssetsWorkBook.pdf

Forest, M., & Pearpoint, J. (1995). *The talking stick reflection.* https://inclusion.com/change-makers-resources-for-inclusion/training-tools/the-talking-stick-reflection

Harrison, R., Blickem, C., Lamb, J., Kirk, S., & Vassilev, I. (2019). Asset-based community development: Narratives, practice, and conditions of possibility—A qualitative study with community practitioners. *SAGE Open, 9*(1), 2158244018823081. https://doi.org/10.1177/2158244018823081

Introduction to Community-Led Co-Design. (n.d.). https://co-design.inclusivedesign.ca/introduction/

Iyer, D. (2022). *Social change now: A guide for reflection and connection.* Thick Press.

Katz, N., & McNulty, K. (1994). *Reflective listening.* Syracuse University. https://www.maxwell.syr.edu/docs/default-source/ektron-files/reflective-listening-neil-katz-and-kevin-mcnulty.pdf?sfvrsn=f1fa6672_7

Kimmerer, R. W. (2022). *Braiding sweetgrass: Indigenous wisdom, scientific knowledge, and the teachings of plants.* Zest Books.

Kretzmann, J., & McKnight, J. P. (1996). Assets-based community development. *National Civic Review, 85*(4), 23–29. https://onlinelibrary.wiley.com/doi/10.1002/ncr.4100850405

Kretzmann, J. P., & Mcknight, J. L. (1993). *Building communities from the inside out: A path toward finding and mobilizing a community's assets.* Northwestern University.

Okun, T. (2021). *White supremacy culture – Still here.* https://www.whitesupremacyculture.info/

Project Implicit. (n.d.). [Assessment]. https://implicit.harvard.edu/implicit/takeatest.html

Russell, C., & McKnight, J. (2022). *The connected community: Discovering the health, wealth, and power of neighborhoods* (1st ed.). Berrett-Koehler Publishers.

Saltmarsh, J. A., & Hartley, M. (2012). *To serve a larger purpose: Engagement for democracy and the transformation of higher education.* Temple University Press.

Sanoff, H. (2011). Multiple views of participatory design. *Focus, 8*(1). https://doi.org/10.15368/focus.2011v8n1.1

Shah, R. W., Troester, J. M. S., Brooke, R., Gatti, L., Thomas, S. L., & Masterson, J. (2018). Fostering eABCD: Asset-based community development in digital service-learning. *Journal of Higher Education Outreach and Engagement, 22*(2), 189–222.

Tunstall, E. (2023). *Decolonizing design: A cultural justice guidebook.* MIT Press.

Yates, J., & Accardi, M. (2019). *Field guide for urban university-community partnerships* (pp. 1–64). Institute for Advanced Studies in Culture University of Virginia. http://iasculture.org/research/publications/thriving-cities-field-guide

2

DEPLOYING DESIGN JUSTICE IN ENVIRONMENTAL INJUSTICE SETTINGS

Lisa L. Phillips

Introduction

Environments change constantly, as do the risks and benefits associated with such change. People respond to these changes through embodied, networked, and mediated ways, shaping how people perceive risk and who bears the brunt of natural and human-designed environmental hazards. While the networked intricacies of global heating and incidents like the Deepwater Horizon disaster associated with a well-design flaw may seem distinct, one informs the other's design by default (Frost, 2013). Donnie Sackey's (2020) call to "embed EJ [environmental justice] principles into our design processes" emphasizes the need for inclusive design across contexts like policymaking, pollution mitigation, and technologies designed to help residents monitor environmental risks (p. 34). I focus less on technology design and more on residents' material engagement with environmental hazards, as they weave sensory rhetoric into EJ initiatives.

To analyze EJ complexity, one must evaluate stakeholders' influences. From policymakers and city planners to lawyers, developers, businesses, and communities, each constituent plays a role in EJ and its inverse, environmental injustice. Sackey (2020) underscores how people in positions of power and privilege design inaccessible technologies, reinforce oppressive biases, and create environmental injustices, making environmental justice a design issue.

Evaluating how people intervene in environmental injustice becomes crucial amid looming large-scale concerns, for tackling environmental injustices one at a time provides insights into wicked problems that involve

DOI: 10.4324/9781003469995-4

overlapped and networked systems, uncertainty, precarity, and value-laden decisions (Blythe et al., 2008; Simmons, 2007). Design theorists Horst Rittel and Melvin Webber (1973) introduced "wicked problems" and highlighted how such problems rebuff efforts to define or solve them. Because wicked problems involve complex and interconnected issues, associated ripple effects obfuscate the problem-solving process. Addressing problems one at a time requires one to understand the bigger picture, yet focus diminishes psychological and political paralysis that can occur when facing complex problems (Williams, 2019).

Environmental injustice refers to how a society inequitably distributes benefits and burdens like exposure to toxic air, water, and soil, and access to natural resources. The concept highlights risks marginalized or minoritized groups face associated with environmental hazards, as elevated exposure rates coincide with factors like race, age, gender, income, or class (Bullard et al., 2007). Environmental injustice manifests in various forms, such as siting polluting industries in low-income neighborhoods, inequitable access to clean air and water, and how governments enforce environmental regulations if such regulations exist. The ongoing Flint, Michigan, lead contamination crisis presents an example.

Environmental injustice accentuates how socio-environmental issues intersect with axes of power and privilege, emphasizing a need for equitable policies and practices. Ergo, environmental *justice* means to treat fairly and involve all people, regardless of race, class, gender, age, ability, or income, in how we advocate for, design, develop, and enforce environmental laws, regulations, and policies. Black EJ scholar and activist Robert Bullard points out how "dominant risk paradigms" like those designed and used by the US EPA and other global and local agencies often fail to account for "subpopulations" and "communities of color" in "various phases of assessing impacts, including scoping, data gathering, discovery of alternatives, analysis, mitigation, and monitoring" (Bullard, 2005, p. 21). Beverly Wright, a Black sociologist and EJ advocate who often coauthors with Bullard, emphasizes EJ within a broader social justice framework. Wright emphasizes how socio-economic inequalities determine proximity to environmental hazards (Wright, 2012, pp. 125–145). Both stress policies and practices designed to foreground EJ like integrated community input, embedded pollution-control policies in building permits, and monitored risk. I focus on practices and policies designed to redress environmental injustice and center majority-minoritized communities' EJ work. The work occurs both in physical and digital spaces, including social media platforms where residents share and design tactics to address EJ concerns.

Design justice (DJ) speaks to EJ because we design better environmental outcomes when we make a concerted effort to do so. For example, in 1974, two researchers Mario Molina and Sherwood Rowland hypothesized that chlorofluorocarbons (CFCs) were ozone-depleting substances that could cause massive damage to Earth's protective sunscreen (Molina & Rowland, 1974). In response, the newly established United Nations Environmental Program (UNEP) created a committee to assess the hypothesis as did CFC manufacturer Du Pont who brought together a team of US CFC industrial producers (Whitesides, 2020, p. 2). US President Jimmy Carter's administration banned CFCs in the US and the National Aeronautics and Space Administration launched a satellite to study the problem (p. 2). Few other nations took up the call for concern during the early 1980s. Global use of CFCs accelerated. In response, the US and UNEP launched more assertive diplomatic talks and tapped into growing public concerns about environmental hazards writ large. The combined impact sparked global interest and led to the Vienna Convention in 1985 when 20 nations acknowledged the risk CFCs posed to ozone depletion (UNEP, 2003). In 1986, government research stations in Antarctica found an "ozone hole" electrifying public outcry (Whitesides, 2020, p. 3). On September 16, 1987, every nation on Earth adopted the Montreal Protocol to eliminate the use and production of chemicals that caused holes in the ozone layer (Mahabadi, 2023). Those holes are healing and the Montreal Protocol represents how people collaboratively designed policy that worked (Gill, 2023). Although scientific data informed the protocol's impetus, it was "non-scientific actors" like diplomats, concerned industry representatives, and the public who shaped "The Little Treaty that Could" into global statecraft (Whitesides, 2020, p. 1). A "diverse coalition" of policymakers, diplomats, scientists, industry, and public pressure resulted in corrective action (p. 8).

Talking back to coalitional action, I provide a four-stage procedural method applied to a case study to listen to how people use sensory rhetoric to evaluate risk and design healthier environments. To parley: can people in minority-majority communities deploy sensory rhetoric to redress environmental hazards, and, if so, how? The rhetorician's answer: yes, they do, to varying degrees of success, usually in coalition. In successful cases, people use their senses to identify environmental problems, participate in vernacular exchanges, engage in mediated public debate, and affect new public policy outcomes or create direct initiatives.

History, culture, politics, and power crease EJ's brow. The chapter irons out how minoritarian communities use sensory rhetoric as a tool to press for effective environmental policy design and decision-making processes. A combined EJ/DJ framework delivers insights technical and professional communication (TPC) scholars, students, and practitioners can use to evaluate inequitable hazard placement and risk assessment realized through embodiment.

A case study of environmental injustice in the McKinley Park neighborhood in Chicago, Illinois, underscores how embodied knowledge and sensation can transform a passive public into a "visceral public" that shapes the mediated public sphere and by extension environmental policy into a spectrum of effects (Johnson, 2016, pp. 2–3). Visceral publics begin with viscera—the gut, nerves, feelings—deeply embodied, intuitive responses beyond rational argument yet inform it. Jenell Johnson characterizes such publics as publics that "emerge from discourse about [bodily] boundaries" and "cohere by means of intense feeling" often around "perceived threats" real, imagined, and gaslit (p. 2). While one can apply the idea of visceral publics to evaluate their negative effects on public policy and partisan politics, like anti-vaxxers, "Trumpeteering," climate change deniers, and so forth, I focus on how visceral publics induce responsive environmental policy through the body's sensational design (Gunn, 2020). Although the chapter does not address the design of monitoring technologies, it offers designers and developers "value sensitive" concepts from EJ/DJ perspectives (Sackey, 2020, p. 39). Such concepts should shape future "wearable" technologies like "personal air monitors" (PAMs) or other devices that amplify human perceptions about environmental conditions (pp. 33–34).

EJ/DJ Framing and Bodystorming

While "design thinking" in TPC should include a focus on "social advocacy while solving technical problems," contending with environmental injustices also calls for an intersectional "bodystorming" approach that involves "emotional perception taking" and making across different environments (Van Boven et al., 2013, p. 119; Verhulsdonk et al., 2021, p. 473). Conceptually like brainstorming, bodystorming relates to design thinking and creative problem solving, as it involves physically acting out or simulating scenarios to address challenges, gain experiential insight, and generate potential solutions to problems posed. Bodystorming allows one to empathize or engage in "emotional perspective taking" of someone else's experience, find pain points, and reveal opportunities for improvement in an embodied manner (Van Boven et al., 2013, pp. 119–120). When it comes to environmental hazards, bodystorming amidst them is a risky proposition, yet it is one that residents in environmental injustice zones do regularly, though they are unlikely to label it thus. Also attuned to lived experience, Phaedra Pezzullo (2009) outlines a method of "toxic tourism" wherein residents invite people to visit a contaminated community to get a better sense of what is involved, what it feels, looks, sounds, or smells like. Neighborhood visitors engage in de facto bodystorming sessions designed to draw more attention to environmental injustices, work to shift attention to local problems that policymakers may otherwise ignore, and create solutions that address the targeted environmental hazards. The bodystorming approach addresses "empathy gaps" common among environmental injustice settings. Empathy gaps are mismatches between how one imagines one would respond to a situation, which differs from their current reality, versus how emotions influence behavior. Empathy gaps appear, in part, because people underestimate how emotions profoundly change them, "if temporarily, shaping their attention, construal, and motivation" (Van Boven et al., 2013, p. 120). For example, someone might think they would be calm and rational when faced with a stressful situation, but when the actual event occurs, their emotions may overwhelm them, leading to a different response than initially anticipated. While emotions may initially overwhelm us, we can also make analytical and deliberative adjustments with these feelings upon reflection (Fountain, 2024, p. 3). Let us visit Chicago's Southwest Side to understand an example.

Exposing and Closing Empathy Gaps

The McKinley Park neighborhood on Chicago's Southwest Side is approximately 55 percent Latino and 30 percent Asian. The MAT Asphalt plant contaminated the community's air and soil for years. Cheryl Newton, the region's US EPA administrator, characterizes McKinley Park as a "community with environmental justice concerns" (quoted in Chase, 2021). In 2023, a Cook County Circuit Court judge approved a $1.2 million settlement in a class-action lawsuit against the plant owners. Residents who live within a half-mile of the facility could file a claim to receive money for their pain and suffering, and the plant must install better pollution controls and mitigation measures designed to improve residents' quality of life and health. This is an EJ success story. Residents' collective grassroots efforts influenced local political leaders who then worked with city officials to design mitigation and improvement efforts. I start with the success story before applying the four-stage process so readers have context on how an EJ/DJ scene unfolded and resulted in a successful outcome.

Initially, city officials dismissed the residents' claims about the asphalt plant's impact on their quality of life and health. An empathy gap grew between city officials and locals. The empathy gap existed among city officials due to their distance from the plant's noxious odors. Imagine the last time you were stuck in a construction zone where steaming asphalt belched out fumes as crews paved the road. Now, imagine living proximal to the stench with no whiff of respite on the horizon. Residents, in contrast to the distanced officials, experienced profound emotional and physical effects related to the odors. Some residents staged local protests in a nearby public park and enacted a bodystorming perspective, inviting regional alderpersons to "nose" the stench for themselves. Immersion in the smellscape influenced policymakers through emotional perspective making that closed key aspects of the empathy gap, which led to deliberation and action.

Olfactible evidence in the rhetorical ecology led to the design of grassroots EJ/DJ efforts and moved policymakers to redress the situation. However, the protracted response time exposed residents to pollution that affected their health and the health of plants, animals, water, air, and soil. Ultimately, residents smelled a stench, invited others to smell it, raised a stink about it in the public sphere, and closed the empathy gap such that a problem met with a solution. The residents would have benefited from affordable wearable PAMs as Sackey has suggested for similar situations (pp. 35–36), but no access existed.

Designing a Four-Stage Procedural Approach to Evaluate Environmental Justice Efforts

TPC scholars have advocated for equitable and democratic ways to rectify environmental concerns (Blythe et al., 2008; Herndl & Brown, 1996; Itchuaqiyaq, 2023; Katz & Miller, 1996; Ross, 2017; Sackey, 2020; Simmons, 2007; Walwema, 2023; Williams, 2023). TPC scholars who study video games draw on procedural rhetoric (PR) to parse virtual environs, and one can apply aspects of PR to analyze and advocate for natural environs, too (Anderson et al., 2018; Bogost, 2007; Holmes, 2017). PR refers to processes or procedures used in interactive systems like video games to form, convey, and evaluate persuasive messages. Understanding a place, space, subject, or object through implicit and embodied means and methods involves PR because interactive systems extend to bodies enmeshed in natural systems. Such bodies include human, nonhuman, and beyond-human sensing bodies in which sensory systems designed by natural and engineered processes convey persuasive rhetorical messages about environmental risk. "Rhetorical ecologies," comingled spheres "of active, historical, [cultural], and lived processes," assemble within frameworks "of *affective ecologies*" that involve emotion, feeling, sensation, and enlighten how we assess risk (Edbauer, 2018, p. 168). In what follows, I articulate a four-stage PR approach to understand how minoritarian communities address environmental concerns using sensory rhetoric.

The procedural method supports us to analyze and evaluate composite sensations, or clusters of sensuous activity, emerging through "visceral publics" experiences with olfactive violence (Johnson, 2016). Olfactive violence emerges from both natural and engineered sources like rotting flesh and chemicals, respectively. Olfactive violence consequently informs mediated artifacts that circulate in the public sphere. Examples of mediated artifacts include social media posts, policy documents, news stories and broadcasts, public meeting minutes, videos, protest signage, and more. Deploying the process helps one evaluate how people make sense of environmental change and makes plain efforts to persuade policymakers to address environmental risks. A procedural approach allows rhetoricians, designers, developers, and others attuned to social justice exigencies to uncover environmental injustices that emerge from olfactive experiences and scent events and shows how olfactory persuasion can prime more sensitive responses to lives elided in public discourse about environmental hazards. Scent "events [are] where experience[s] actualize," and where nonhuman and human concerns merge (Manning, 2016, p. 3). Shifts in the smellscape compose upon the body an immersive environmental experience. Deploying the process I introduce can also support the city and rural planners to identify areas where PAMs could prove useful to residents. Cumulative effects of olfactive violence create lasting physical and mental impacts even if the toxic chemical threshold is low (Bushdid et al., 2014; Calderón-Garcidueñas et al., 2010; Cerulo, 2018; Croy & Hummel, 2017; Hinton et al., 2004).

Olfactory persuasion—how we perceive and discuss what we smell—helps us understand, evaluate, and redress specific forms of environmental injustice. As we interact with different environments, our senses help us navigate those environments and assess whether they are safe. Digital sniffing tools like PAMs augment our senses and can measure air contaminants. Often, the tools receive more credence than our natural senses, particularly in legal contexts where dominant risk assessment paradigms involving test tubes, labs, science, and reason are privileged, yet people usually note the impact of environmental hazards and risks on and through bodily sensations that are "physical … tangible and felt" and informed by prior experiences and perceptions (DeVasto et al., 2016, p. 139). Our sense of smell has a unique intimacy with our emotions and memories connected to our flight-or-fight responses in the brain. Odors violate boundaries, defy containment, and index the material presence of other entities in our midst.

Stage One—Sensing Emergent Problems

In stage one, an environmental problem must be sensible, sensed, or readily perceived. Humans, nonhuman entities like dogs, birds, plants, and insects, or machine sensing like PAMs, e-noses, pressure sensors, or other apparatus constitute stage one sensors. For example, trained dogs and rats can detect bombs and other risks. While the scientific debate is unresolved, some animals and insects sense subtle shifts in the earth's crust, or fore-shocks, prior to an earthquake (Quain, 2021). Sailing aficionados watch barometers and scrutinize weather forecasts, but seeing shorebirds head inland reliably indicates approaching storms, as the birds accurately "read" barometric pressure with their bodies. In stage one, humans may hear, see, smell, taste, or feel something amiss in the environment and then find the source to assess the threat posed. Targeted study of physiological sensation present in human and nonhuman living entities provides a good base for the four-stage process, as no expensive equipment is involved and sensation directs attention to environmental changes.

McKinley Park residents have gone through the four-stage process several times in their efforts to address environmental injustices. McKinley Park is a working-class neighborhood that historically has been the site of steel mills, foundries, and other industries notorious for hazardous waste production and "good," albeit dangerous, industrial jobs (Mack, 2015). Matching the rise of the Industrial Revolution, people filled in the "Ducktown" marsh with ash, garbage, and all sorts of flotsam to abate the swampy conditions in the mid-to-late 1800s. After the Great Chicago fire of 1871, some industries moved to unharmed McKinley Park (pp. 43, 47). Within five years of the fire, 11 steel and ironworks and 27 brick-yard factories opened. Meatpacking plants to the immediate south went full boar and the offal ended up in surrounding sloughs. Upton Sinclair infamously labeled the result "Bubbly Creek" in *The Jungle* (1906, p. 39). The associated stench was gut-wrenching and pervasive. The meatpacking industry was notorious for both labor exploitation and environmental impact. In the early 1900s, as modes of and for transportation became easier, McKinley Park and the surrounding area became part of a manufacturing district that included Pepsi-Cola, the Wrigley Company, and the Chicago-Sun Times publishing and distribution plant among other smaller industrial facilities, most of which were unregulated. In the mid-1920s, after years of complaints about the lack of green space, the city built a public park and named it after slain US President McKinley. The neighborhood now has the same name, replacing "Ducktown."

The background matters, as community history and memories inform how residents react when new industries replicate the same kinds of injuries. In 2017, McKinley Park residents formed the group "Neighbors for Environmental Justice" (N4EJ) in response to a hot-mix asphalt plant due to noxious odors and other effluent associated with the plant. The nonprofit group includes residents from a range of backgrounds, and they network with adjacent communities who have similar EJ concerns. Affiliate groups include the Little Village Environmental Justice organization founded in 1994 to contend with coal-fired power plants associated with cancer clusters and respiratory maladies (Bates, 2020). N4EJ has an active social media presence across a range of platforms and multilingual outreach matters (Neighbors for Environmental Justice [N4EJ], n.d.). N4EJ also assists residents with air quality initiatives that include enforcing pollution compliance and supports residents' efforts to report sensory-based air quality data. For example, a Facebook post on May 22, 2019, specifically directs residents on how to report odor complaints to the EPA and to a non-emergency 311 phone number. N4JE members provide English and Spanish instructions on a linked webpage to help people document incidents and file formal complaints. Access both the English and Spanish infographics here: https://www.facebook.com/photo/?fbid=26092853524 32568&set=pcb.2609285465765890

Stage-one scent events shift with the wind direction, the heat of the day, and the pollution produced. Residents' exposures to airborne toxins and olfactive violence overlap. Of the stage one experiences, one resident reported, "I have to kind of sniff outside, does it smell clean yet? ... I don't want [my child] to be outside if it's very strong, so it controls her day. We don't know what exactly is in it" (N4EJ, n.d.). Another resident reported a different event, "[I] took just a few steps [outside] and was engulfed with overwhelming #toxic odors.... I became concerned for my health, and to be honest, I was livid" (N4EJ, n.d.).

Stage Two—Articulating Group-Based Sensory Experiences to Publics

In the second stage, people sense an environmental problem and discuss it afterward with others. If someone, something, or a multitude, senses something amiss within an environment or social scene and does *not* communicate sensual experience, then stage one events fail to appear in the public sphere or circulate in the form of mediated artifacts. The reverse is also accurate. When people do communicate sensual experiences, then composite stage one sensations appear in the public sphere and circulate in the form of mediated artifacts. If people choose not to communicate about unpleasant sensory experiences, it is often related to economic interests like property value concerns, time to converse, access to the mediated public sphere, and other issues tied to capitalism or hegemonic ideologies.

Returning to the McKinley Park example, residents who experience a stage one scent event report their experiences to others in stage two communication in social media and in person and help one another determine how and to whom they should report their concerns. In other words, they informally collaborate and often use olfactory rhetoric. Two

Facebook posts from the group from 2021 and one from 2020, provide salient illustrations:

[I] woke up at 3a on Tuesday morning to my toddler sporadically coughing over the baby monitor to realize my bedroom and whole house was saturated in the smell of MAT Asphalt. Very strong odor nuisance inside of my bedroom, kitchen, and bathroom. I remained awake for about an hour listening to my toddler continue to cough while the odor persisted. It was the worst and most invasive odor nuisance of 2021 so far.

Once again McKinley Park residents who left their windows open overnight to enjoy the cool [September 2021] weather woke to the smell of asphalt in their homes. And remember – that air monitor next to the plant showing the obvious increase in particulate matter when they started operating? It's on a school: the National Latino Education Institute.

Odors coming from Mat Asphalt are unbearable! My family is feeling sick with headaches if we go outside. It smells like burning petroleum and there is a cloudy haze in the air. We believe they have increased production because they will have to close soon for winter. The smells come inside our house even with the windows closed. I have asthma and this is unfair and dangerous to our health.

The olfactory rhetoric McKinley Park residents engage in transmits personal sensory experiences to the group, and residents often share data to corroborate their lived experiences using scientific data to warrant embodied claims, as the second passage above illustrates. Another resident re-posted a tweet from the SW Side Air Network (SWSAN) group to notify others of air quality index alerts issued for the neighborhood. The air quality alert warns of "health effects if ... exposed for 24 hours" and tells "members of sensitive groups" they "may experience more serious health effects with 24 hours of exposure" (SWSAN, n.d.). In response, another poster writes, "This is awful, I would imagine they start heating that stuff [asphalt] in the middle of the night every day before they open for business" (N4EJ, n.d.). Importantly, humans have minimal olfactory perception during deep sleep (Carskadon & Herz, 2004), yet people still breathe in contaminants, so the asphalt plant owners knowingly upped production in the middle of the night. Yet another declares, "It feels like we're secluded from the rest of Chicago on this issue. If you're not going through it every day, you'd don't have to take ownership over the issue that affects the next community over" (N4EJ, n.d.). The poster exposes an empathy gap that stage three takes up.

Stage Three—A Visceral Public Broadcasts Sensory Experiences to Elicit Action

In stage three, a visceral public acts on an environmental problem and uses the mediated public sphere to publicize the issue. Stage three communication differs from stage two due to broader reach and scope. People need to have focused energy to address an environmental injustice problem, and that involves affect and emotion, time, and access to resources. Stage three is where we may see mediated responses to composite sensations fizzle out because those with the power to address an emergent environmental problem may ignore or suppress the situation out of ignorance, an empathy gap, or for a host of unethical reasons (e.g., financial, racist, sexist, ableist, and classist).

Back to McKinley Park and stage three human response, on September 14, 2021, a resident's social media post notes "canceled" 311 complaints. An unidentified public official canceled the odor complaints "due to reason of **Other**," which incensed residents and led to additional stage three communications. Community members speculated why the complaints received summary dismissal, collaborated on a letter to elected Alderman Cardenas, and included a video clip illustrating visible evidence of the pollution cloud over the neighborhood. To officially complain to state and local environmental protection and regulatory agencies in the Chicago area requires residents to log in to an online portal containing a form designed to capture the location, intensity, duration, and description of an odor and its embodied impact. Residents can also call and report the data via 311. In the letter re-posted on social media, residents wanted to know why officials ignored their complaints, what inspection protocols were in place, and what the city intended to do about the persistent problems. Residents asked about the design process in place to address environmental concerns. An independent neighborhood-focused newspaper, *Block Club Chicago*, ran a story about air quality concerns and a public meeting notice for a "People's Hearing," a public protest hosted by residents in the park named after McKinley (Asimov, 2023). Though this predates the 2021 posts, it coincided with the height of the COVID-19 pandemic, and residents wearing masks and signs appear just under the headline. *The Chicago Tribune* and *Chicago Sun Times* later picked up the story, which added pressure to politicians and policymakers who then addressed the issue in stage four.

Stage Four—Broadcasted Sensations Transform Policy

Finally, in stage four, a visceral public's broadcasted and circulated sensations must transform or move policymakers to redress the concerns, which involves intersecting axes of power that affect EJ issues. McKinley

Park had a decisive, positive stage four outcome. The community designed effective rhetorical messaging and launched a coordinated effort informed by N4EJ and their affiliated South Side Chicago-based partners like the Little Village Environmental Justice group, People for Community Recovery, Southwest Environmental Alliance, and the Chicago Environmental Justice Network. A bodystorming visceral public present and active in McKinley Park combined with olfactory rhetoric worked to persuade public officials to resolve the health complaints that stemmed from the asphalt plant's obnoxious scent events. That Chicago's EJ communities are politically engaged and active helps inform why the use of sensory rhetoric in this example aided in the design of more effective pollution controls. US Senators Dick Durbin and Tammy Duckworth, both Democrats, also prompted President Biden's EPA to investigate and resolve the issue (Chase, 2021). Local laws and political landscapes affect response to odor nuisances. Environmental Justice advocates Robert Bullard and Beverly Wright (2012) explain how and why EJ communities need to design anticipatory laws and policies to circumvent problems before they arise. In McKinley Park, the N4EJ supports residents and local government agents to build anti-nuisance clauses and preventive measures into any new business development or expansion plans for existing industry settings. Such work built upon vexed phronesis, however, does little to help communities address pre-existing racist siting policies that put majority-minoritized communities in harm's way from the outset. What redesign ought you and I consider relative to such a wicked problem? Daniel Wildcat (Yuchi Muscogee) instructs, "The most difficult changes required are not those of a physical, material, or technological character, but changes in worldviews and the taken-for-granted values and beliefs that are embedded in modern, Western-influenced societies" (2009, p. 5). Changes in worldview adjust how people understand themselves in "the web of life" in which interdependency as a matter of course ripples across landscapes. Designing with such interdependency in mind is paramount.

Key Takeaways and Design Activities

Under the EJ/DJ framework, the four-stage process could prove useful to TPC, developers (of both land and technologies), and communities to evaluate how, when, where, why, and to what effects people use sensory rhetorics to redress emergent environmental problems. The procedural approach provides a practical method we can use to find interdependent environmental issues, track visceral publics' sensational efforts, look for mediated artifacts that circulate in response, and evaluate policy design and the effects on the web of life. Reducing the empathy gap, the process can also assist policymakers or risk communicators to understand when, where, and why sensory rhetoric might not suffice as evidence of environmental risk and how to address community health concerns respectfully, ethically, and effectively. Human sensory rhetoric can co-mingle with nonhuman, machine sensing, and social media technologies to triangulate data in EJ zones and in legal cases, as Sackey aptly notes (2020). A bodystorming process could help inform the design and outcomes of anticipatory language built into permitting processes to prevent different forms of environmental hazards from taking place or gaining traction due to inadequate policies. As an illustration, if we apply the procedural method to multiple sites across a geographical locale, then patterns may emerge tied to the region's dominant ideologies, power structures, and politics. Elsewhere, I address how rural and politically conservative communities deploy olfactory persuasion to address environmental issues. Knowledge of the patterns will lead to more informed arguments designed to redress environmental injustices. Imagine how an EJ community could include language in new building permits to address odor nuisances, eyesores, excessive noise, or other environmental impacts associated with diverse types of businesses the procedural process reveals as concerns. Finally, different grassroots EJ groups might find the process useful to understand when/ if during the four-stage process communication breaks down or when to merge human sensory responses with machine sensing technologies when qualitative evidence calls for more support to help a community address health and other socio-economic and environmental concerns.

Design activities associated with the four-stage process for pedagogical purposes could include:

- Develop a bodystorming activity that focuses on a specific site or sense to support students' or practitioners' capacities to empathize with users in a given space or place;
- Apply the four-stage process to different environments both in physical and virtual spaces to evaluate problematic design or development from a sensory rhetorics perspective;
- Create purposeful empathy gaps and interrupt them with bodystorming approaches based on different sense perceptions to evaluate accessibility beyond vision and hearing;
- DJ/EJ design initiatives could include mapping and ArcGIS story mapping as community-engaged projects (Stephens & Richards, 2020);
- Conduct smellwalks when considering urban planning or architecture to understand where problem areas exist and why.

References

Anderson, B. R., Karzmark, C. R., & Wardrip-Fruin, N. (2018). The psychological reality of procedural rhetoric. *Proceedings of the 14th International Conference on the Foundations of Digital Games. Association for Computing Machinery, 44*, 1–9. https://doi.org/10.1145/3337722.3337751

Asimov, N. (2023, December 13). MAT Asphalt agrees to $1.2 million class-act settlement with McKinley Park neighbors. *Block Club Chicago.* https://blockclubchicago.org/2023/12/13/controversial-mat-asphalt-reaches-1-2-million-class-action-settlement-with-mckinley-park-neighbors/?

Bates, J. C. (2020). Local expertise, global effects: Amplifying activist arguments for climate change action. *Enculturation: A Journal of Rhetoric, Writing, and Culture, 32*. https://enculturation.net/Local_Expertise_Global_Effects.

Blythe, S., Grabill, J. T., & Riley, K. (2008). Action research and wicked environmental problems: Exploring appropriate roles for researchers in professional communication. *Journal of Business and Technical Communication, 22*(3), 272–298.

Bogost, I. (2007). *Persuasive games: The expressive power of video games.* MIT Press.

Bullard, R. D. (Ed.). (2005). *The quest for environmental justice: Human rights and the politics of pollution.* Sierra Club Books.

Bullard, R. D., Mohai, P., Saha, R., & Wright, B. (2007). *Toxic wastes and race at 20, 1987–2007: Grassroots struggles to dismantle environmental racism in the United States.* Report prepared by the United Church of Christ Justice and Witness Ministries.

Bullard, R. D., & Wright, B. (2012). *The wrong complexion for protection: How the government response to disaster endangers African American communities.* New York University Press.

Bushdid, C., Magnasco, M. O., Vosshall, L. B., & Keller, A. (2014). Humans can discriminate more than 1 trillion olfactory stimuli. *Science (American Association for the Advancement of Science), 343*(6177), 1370–1372. https://doi.org/10.1126/science.1249168

Calderón-Garcidueñas, L., Franco-Lira, M., Henríquez-Roldán, C., Osnaya, N., González-Maciel, A., Reynoso-Robles, R., & Doty, R. L. (2010). Urban air pollution: Influences on olfactory function and pathology on exposed children and young adults. *Experimental Toxicology Pathology, 62*, 91–102.

Carskadon, M. A., & Herz, R. S. (2004). Minimal olfactory perception during sleep: Why odor alarms will not work for humans. *Sleep, 27*(3), 402–405.

Cerulo, K. A. (2018). Scents and sensibility: Olfaction, sense-making, and meaning attribution. *American Sociological Review, 83*(2), 361–389. https://doi-org.lib-e2.lib.ttu.edu/10.1177/0003122418759679

Chase, B. (2021, July 28). Feds jump into McKinley Park asphalt plant fracas. *Chicago Sun Times.* https://chicago.suntimes.com/2021/7/28/22598956/mat-asphalt-mckinley-park-epa-environmental-protection-dick-durbin-tammy-duckworth-michael-tadin?fbclid=IwAR3WSgMwHA4SdX8NEYKPGxjjBjUqf Tt3j6zA74Q0udKHcCVazgKFiey1jZ4

Croy, I., & Hummel, T. (2017). Olfaction as a marker for depression. *Journal of Neurology, 264*, 631–638. https://doi.org/10.1007/s00415-016-8227-8

DeVasto, D. et al. (2016). Statis and matters of concern: The conviction of the L'Aquila seven. *Journal of Business and Technical Communication, 20*(2), 131–164.

Edbauer, J. (2018). Unframing models of public distribution: From rhetorical situation to rhetorical ecologies. In J. Gunn, & D. Davis (Eds.), *Fifty years of rhetoric society quarterly: Selected readings, 1968–2018* (pp. 165–182). Routledge.

Fountain, T. K. (2024). Rhetoric and the cultural politics of Donald Trump. *Rhetoric Society Quarterly, 54*(1), 1–12. https://doi.org/10.1080/02773945.2023.2295772

Frost, E. A. (2013). Transcultural risk communication on Dauphine Island. *Technical Communication Quarterly, 22*(1), 50–66. https://doi.org/10.1080/10572252.2013.726483

Gill, A. (2023). Lessons from Montréal: Creating MNC support for environmental regulations. *Flux: International Relations Review, 14*(1), 121–129. https://doi.org/10.26443/firr.v14i1.155

Gunn, J. (2020). *Political perversion: Rhetorical aberration in the time of Trumpeteering.* University of Chicago Press.

Herndl, C. G., & Brown, S. C. (Eds.). (1996). *Green culture: Environmental rhetoric in contemporary America.* University of Wisconsin Press.

Hinton, D., Pich, V., Chhean, D., & Pollack, M. (2004). Olfactory-triggered panic attacks among the Khmer refugees: A contextual approach. *Transcultural Psychiatry, 41*(2), 155–199. https://doi.org/10.1177/1363461504043564

Holmes, S. (2017). *The rhetoric of videogames as embodied practice.* Routledge.

Itchuaqiyaq, C. U. (2023). When the sound is frozen: Extracting climate data from Inuit narratives. In S. D. Williams (Ed.), *Technical communication for environmental action* (pp. 19–38). State University of New York Press.

Johnson, J. (2016). "A man's mouth is his castle": The midcentury fluoridation controversy and the visceral public. *Quarterly Journal of Speech, 102*(1), 1–20.

Katz, S. B., & Miller, C. (1996). The low-level radioactive waste siting controversy in North Carolina: Toward a rhetorical model of risk communication. In C. G. Herndl, & S. C. Brown (Eds.), *Green culture: Environmental rhetoric in contemporary America* (pp. 111–140). University of Wisconsin Press.

Mack, A. (2015). *Sensing Chicago: Noisemakers, strikebreakers, and muckrakers.* University of Illinois Press.

Mahabadi, D. (2023). Enhancing fairness in the Paris Agreement: Lessons from the Montreal and Kyoto protocols and the path ahead. *International Journal of Environment and Sustainable Development, 22*(3), 320–348. https://doi.org/10.1504/IJESD.2023.132088

Manning, E. (2016). *The minor gesture.* Duke University Press.

Molina, M. J., & Rowland, F. S. (1974). Stratospheric sink for chlorofluoromethanes: Chlorine atom-catalysed destruction of ozone. *Nature, 249*(5460), 810–812.

Neighbors for Environmental Justice. [Facebook page]. (n.d.) Retrieved February 2, 2024, from https://www.facebook.com/n4ejchicago/

Pezzullo, P. (2009). *Toxic tourism: Rhetorics of pollution, travel, and environmental justice.* University of Alabama Press.

Quain, A. (2021, September 22). Can animals sense when an earthquake is about to happen? *The Conversation.* https://theconversation.com/can-animals-sense-when-an-earthquake-is-about-to-happen-168483

Rittel, H., & Webber, M. (1973). Dilemmas in a general theory of planning. *Policy Sciences, 4*(2), 155–169.

Ross, D. G. (Ed.). (2017). *Topic-driven environmental rhetoric.* Routledge.

Sackey, D. J. (2020). One-size-fits none: A heuristic for proactive value sensitive environmental design. *Technical Communication Quarterly, 29*(1), 33–48. https://doi.org/10.1080/10572252.2019.1634767

Simmons, W. M. (2007). *Participation and power: Civic discourse in environmental policy decisions*. State University of New York Press.

Sinclair, U. [1906]. (2016). *The jungle*. Clydesdale.

Stephens, S., & Richards, D. (2020). Storymapping and sea level rise: Listening to global risks at street level. *Communication Design Quarterly*, 8(1), 5–18. https://dl.acm.org/doi/abs/10.1145/3375134.3375135

SW Side Air Network. [Twitter post]. (n.d.). Retrieved February 2, 2024, from https://www.facebook.com/n4ejchicago/photos/a.2124880700873038/2982984995062600

United Nations Environmental Programme. (2003). *Handbook for the international treaties for the protection of the ozone layer: The Vienna Convention (1985), the Montreal Protocol (1975)*. UNEP/Earthprint.

Van Boven, L., Loewenstein, G., Dunning, D., & Nordgren, L. F. (2013). Changing places: A dual judgment model of empathy gaps in emotional perspective taking. *Advances in Experimental Social Psychology*, 48, 117–171. https://doi.org/10.1016/B978-0-12-407188-9.00003-X

Verhulsdonk, G., Howard, T., & Tham, J. (2021). Investigating the impact of design thinking, content strategy, and artificial intelligence: A "streams" approach for technical communication and user experience. *Journal of Technical Writing and Communication*, 5(4), 468–492. https://doi.org/10.1177/00472816211041951

Walwema, J. (2023). Participatory policy enacting technical communication for a shared water future. In S. D. Williams (Ed.), *Technical communication for environmental action* (pp. 245–266). State University of New York Press.

Whitesides, G. (2020). Learning from success: Lessons in science and diplomacy from the Montreal Protocol. *Science & Diplomacy*, 9(2), 1–13.

Wildcat, D. R. (2009). *Red alert!: Saving the planet with indigenous knowledge*. Fulcrum.

Williams, C. M. (2019). Analysis of factors related to science teachers' perceptions about climate change: Implications for educators. [Doctoral dissertation, Texas Tech University]. ProQuest. https://hdl.handle.net/2346/85313

Williams, S. D. (Ed.). (2023). *Technical communication for environmental action*. State University of New York Press.

Wright, B. (2012). Race, politics and pollution: Environmental justice in the Mississippi River chemical corridor. In R. D. Bullard, J. Agyeman, & B. Evans (Eds.), *Just sustainabilities: Development in an unequal world* (pp. 125–145). Taylor and Francis Group.

3

TRUST, UNDERSTAND, ACT

Using Visual Place-Based Research Methods to More Deeply Understand Community Perspectives

Erin Brock Carlson

Introduction

Those of us doing community-engaged work in technical and professional communication (TPC) are continually seeking out ways to build meaningful relationships with community members. Scholars in the field have developed a range of heuristics that support this kind of work (Gonzales, 2022; Rose & Cardinal, 2021; Walton et al., 2019) that urge us to center justice in our work. These heuristics can guide us as we develop and navigate community-engaged projects, but equally important are the research methods we use in such projects. Research methods are a bridge between the heuristics that guide us and the actual needs of the communities we serve.

Place-based research methods, which often incorporate visuals, provide opportunities to document complex problems and how those problems affect everyday lives. Images are a powerful tool for capturing complexity because they can depict multiple factors at once. Contextualizing those images relative to cultural or geographical location builds on that power by highlighting how problems are experienced daily *and* to encourage possible solutions that respond to real community needs. Stakeholder-created depictions of problems that explicitly address the role of the cultural and geographic place offer a meaningful acknowledgment of how financial, environmental, and social inequities are often unnoticed or undocumented; research approaches that explicitly seek out stakeholder-created depictions of problems build relationships between community members and encourage collaborative problem-solving.

DOI: 10.4324/9781003469995-5

This chapter begins with a brief overview of how underrepresented communities are oftentimes underrepresented in multiple ways. Specifically, I offer rural areas as an example of this dynamic, as they are framed in one-dimensional ways that obfuscate the lived realities of multiply marginalized residents within rural communities. I then discuss visual-based methods, specifically those that incorporate a focus on place, to document public problems and how they are experienced by underrepresented groups. I provide brief overviews of three methods I have used in such work: participant-generated imagery (PGI), go-along site interviews supplemented by photography, and collage-making. Next, I outline three affordances of visual place-based methods that emphasize their flexibility and responsiveness to community concerns. I conclude with an outline for a place-based problems portfolio activity that can support the development of a shared vision for community action.

Underrepresented Communities, Undocumented Perspectives

Before diving more deeply into the possibilities of visual place-based research methods specifically, I would like to discuss the power of community-engaged research approaches in communities with underserved populations. Jones and Walton (2018) urge others to consider "how communication broadly defined can amplify the agency of oppressed people – those who are materially, socially, politically, and/or economically under-resourced" (p. 242). Community-based research methods support this goal by providing community members whose concerns often go overlooked opportunities to articulate their experiences in ways that resonate with them. Because visual methods allow for a range of types of evidence that are generated by participants (in comparison to, for example, answering specific interview questions designed by researchers), multiple marginalized individuals and communities have more control over their responses in research settings, resulting in greater levels of engagement.

Though marginalization is often attributed to individual identities (for example, a cis woman might link an experience to her identity as a cis woman and histories of oppression related explicitly to gender), marginalization often occurs at the intersection of multiple identity factors. Theories of intersectionality (Crenshaw, 1989) acknowledge the layering of identities and subsequent injustices, noting how certain communities are multiply marginalized and "experience oppression through their positions at the edges of societal and organizational decisions, cultural representation, and legitimated experience and expertise" (Walton et al., 2019, p. 19). A range of factors contribute to lived experiences though each factor is often understood as isolated despite this layering.

One aspect of identity that benefits from an intersectional approach is place, specifically, rural places. Since rural communities are largely represented as homogenous, diverse members of those communities are oftentimes ignored; LGBTQ+ people (see Nichols, 2013), Native Americans and indigenous communities (Reddy, 2019), Black residents (Spriggs & Paden, 2018), and migrant workers (Lippard & Spann, 2014) are all underrepresented in narratives about rural communities in the United States. Because rural communities are often painted with broad strokes by media and research alike, the perspectives of diverse residents in largely rural regions like Appalachia (where my research and teaching take place) can be difficult to find. This lack of representation, combined with histories of exploitation, can contribute to suspicion toward community-based research, especially when researchers are new to the area or have not yet built relationships with key stakeholders within the community. As a result, research approaches that prioritize place and use visual approaches are useful, as they embrace collaboration and position community members as experts in their own right.

Furthermore, research approaches that explicitly strive for greater levels of inclusion by amplifying community perspectives that often go overlooked can serve social justice causes. Young (1990) offers one definition of justice as: "the institutionalized conditions that make it possible for all to learn and use satisfying skills in socially recognized settings, to participate in decision-making, and to express their feelings, experience, and perspective on social life in contexts where others can listen" (p. 91). For Young, the capacity to share experiences in approachable and even empowering ways is inextricably linked to meaningful participation in public life. Sharing results in increased levels of empathy, which is a foundational attribute of several approaches to justice-oriented research, including design thinking. Visual place-based methods are one approach to expanding the modes of participation in community-based research, and therefore, an approach to addressing complex public problems in just ways.

Visual Methods and Their Contributions to Community-Engaged, Place-Focused Research

Visual research methods have received increased attention in recent years due to their flexibility and the ease with which certain methods can be used thanks to advancements in digital technology. In a field already concerned with the role of visuals in communication, researchers in TPC have taken up visual methods and applied them in a range of contexts: visual ethnography to document design processes (Weedon, 2017); "living visual voice" with seniors (Swacha, 2021); collage, drawing, and mapping with Rwandan youth (Walton et al., 2015); visual frame analysis to examine xenophobic and racist depictions of photos in news articles (Batova, 2021); "chorography" using layered mapping (Gogan & Scott, 2023), and more. Many of these methods, because of their focus on interpretation, invite participation from research participants, leading to moments that "may help researchers and designers develop rich intersubjective understandings and representations of communication design contexts" (McNely, 2013, p. 126). That is, visual methods can oftentimes illuminate multiple concerns at once, such as the role of the body or environment in experiencing the world.

In this chapter, I describe three visual research methods that lend themselves to a focus on place. Though these are just three of many possible approaches, they do exemplify the types of meaningful collaboration that can emerge from the use of visual place-based methods that position researchers and community members as co-interpreters of data.

Participant-Generated Imagery

PGI, sometimes referred to as photovoice, is a research method that asks participants to take photographs in response to particular prompts. Originating in public health and now popular in many disciplines, the method is designed so that participants might "record and reflect their community's strengths and concerns," further conversations about community concerns, and ideally foster policy changes (Wang & Burris, 1997, p. 370). By taking photographs of their daily lives, participants can document their experiences and then later reflect on those experiences in interviews or focus groups where they share their pictures.

I have used PGI in several community-based projects, including a year-long project with a group of community organizers located across the Appalachian region of the United States (Purdue IRB Protocol #1804020503). Each of them was dedicated to social justice concerns, serving different needs including programming for Black youth, environmental justice projects, and problems related to socioeconomic inequality like affordable housing and food security. Together, we came up with general prompts that would remind participants of our shared concerns (place, community need, and social change) and organizers took pictures responding to the prompts over the course of a year. They also wrote short reflections on their experiences to supplement the photos, and we held three focus groups to collaboratively reflect on similarities and differences across their experiences. Together, we assembled a narrative about organizing and social change across the Appalachian region with rich examples of community-based problem solving, and in the process, participants built meaningful connections with one another as the project progressed over time. (See Carlson, 2020, for more on the design of this project.)

Go-Along Interviews and Photographs

Go-along interviews are simply interviews where researchers accompany participants as they travel through their surroundings; however, rather than that travel being merely an aspect of the interview, the movement through an environment is the point of the interview itself. Go-along interviews allow the researcher to be "walked through people's lived experiences of the neighborhood" (Carpiano, 2009, p. 264). Go-along interviews are particularly useful when investigating issues related to place. This method was introduced to me through a collaborative research project with a colleague in Geography whose expertise lies in (unsurprisingly) place. We were working with a regional nonprofit dedicated to environmental justice and designed a study that would document the experiences of people in West Virginia, Ohio, and Pennsylvania living close to gas pipeline infrastructure (West Virginia University IRB Protocol #2005994768). We interviewed participants at their homes in addition to asking them to take photographs of pipeline development and its impact on their daily lives.

During these interviews, participants would take us on tours around their property to show us the actual landscape and how the infrastructure was positioned, often supplemented by photos they had previously taken before we even contacted them. Photographs were a helpful supplement to these go-alongs since participants could actually show us how development had changed the landscape over time, demonstrating that visual methods can be layered upon one another to offer participants multiple ways of telling their stories. (See Carlson & Caretta, 2023, for more on how photos were incorporated throughout the study design.)

Collage Making

Arts-based research approaches describe a range of methods, including collage making. One major advantage of collage is that it has a relatively low barrier to entry; that is, regardless of one's creative background or skill, using materials such as pictures, artifacts, words, paper, and related objects can foster creative expression in ways that a blank page might not: "Collage may jar us into new insights, tear apart and reconfigure ideas, and rework old patterns of thought" (Chilton & Scotti, 2014, p. 170). Participants' reimagination of texts allow them to similarly reconsider their own perceptions and ideas.

Collage making was central to a project I worked on with graduate students and an employee of a statewide nonprofit organization dedicated to the arts (West Virginia University IRB Protocol #2205580033). Our project's purpose was to reach out to artists from multiple marginalized communities across West Virginia through listening sessions, including BIPOC artists, queer artists, disabled artists, and those in addiction and recovery circles. Collage was a natural choice for session facilitation because it was a way for artists of different backgrounds to create and connect with each other. We asked participants to use materials we provided or that they brought themselves to create collages representing their past, present, and future as artists in West Virginia. Participants reflected on both the process and the product of the sessions in relation to larger concerns, leading to meaningful conversations amongst groups. (See Wertz et al., 2023, for more on the origins and development of this project.)

Three Affordances of Visual, Place-Based Methods in Community-Based Projects

One of the most powerful aspects of using visual methods in place-based projects is that they bring participants into the interpretive work of the project. Visual methods do this in two main ways: first, creation prompts for visuals are typically more general in order to offer participants a range of possibilities for their contributions; and second, as visuals are texts inherently subject to interpretation, participants might find vastly different meanings in the same visuals. Such methods, then, offer participants the opportunity to document their environments and experiences in ways that make sense to them, leading to articulations of problems that pinpoint exactly how those problems impact their daily lives, contextualization of those problems in the shared identify factor of place, and connections to overarching conversations about equity and justice.

Affordance One: Building Trust Amongst Project Stakeholders

One outcome of community-based research methods is an increase in trust between researchers and participants. Community-based projects often take an extended period of time to design, having multiple stages in which ideally feedback is sought from participants as the project progresses. Some even invite participants to co-design aspects of the project, which positions participants in a co-researcher role. For projects explicitly focused on place, visual methods can easily invite discussions that emphasize participants' intimate knowledge of their cultural and geographical location, boosting their engagement in the project.

For example, my project with Appalachian community organizers involved them at multiple stages in project design. The first time we met, we had a 3-hour long session where we talked about participatory research methods and PGI, discussed the role of research and how it might enhance their work, and crafted research questions and photo/reflection prompts to guide the project. Speaking frankly about their hopes (and in some cases, skepticism) about academic research helped participants and myself get on the same page while simultaneously creating mechanisms for check-ins. Several participants shared that they felt more comfortable sharing their experiences through writing, rather than photographs. As we talked, we realized that based on their own preferences and comfort levels, they could lean more heavily toward producing photographs or writing short reflections about their experiences. When we met in focus groups periodically, they could share either. What we found was that even those participants who preferred writing to photographing still took photographs that captured their experiences: they had fewer, but still enough to share and collaboratively interpret during our group sessions. And because they were able to participate in a way that they felt comfortable with, they felt ownership of their contributions and trusted me more because I had listened to their needs.

Feelings of trust and shared ownership often increase amongst participants as well, as they meet throughout a project and share their experiences. Over the course of the project with community organizers, we held three focus groups: one two months into the project, one at six months, and one at 12 months. As a result, these conversations changed over time. The first focus group involved a lot of reporting back: participants sharing photos of their work and everyone asking questions about their specific projects. The second focus group was a mix of photographs and larger reflections, with more of a focus on identifying patterns across participant photos, resulting in more metacommentary about the project. The final focus group was the most informal; we barely looked at photographs, and instead, the conversation was very relational, with folks eager to share their overarching thoughts on social justice work in rural Appalachia. This deepening of reflection—from simply documenting their daily tasks to commenting on the power imbalance between nonprofit workers and overarching organizations—was a direct result of an increased level of trust amongst participants that was at least partially fueled by our research. Table 3.1 shows excerpts from each focus group that capture the style of conversation that unfolded.

Each of these excerpts comes from the first five minutes of each focus group. In the first focus group, participants shared their photos and categories. In the second, participants referenced their photos as overall

TABLE 3.1 Excerpts from focus groups showing change in conversation topic over time.

	Initial comment	Response from another participant
Focus Group 1	Um, so I came up with about four different categories, and the first one is capitalists, the next one is events/community forums, the next one is Black neighborhoods or black environments...and the other one is new opportunities for me.	So, I did take a lot of photos, and as I was uploading them for my own sake of trying to figure out where to place them and keep track, I had categorized them ahead of time.
Focus Group 2	I feel like mine was kind of similar to K's and other people's where I'm in the same places a lot and the pictures that—I literally made a list of pictures I would have taken because I've been really bad about taking pictures. But I think, one thing that I've noticed has changed within, like, my project is, I'm running into more young people in the community.	I think kind of similar to what other people are saying, I feel like I've always tried to not take pictures of the same thing and so by nature of that I have not that many pictures. I have more than the last time, I think—but they're of totally different things and I don't know, they're just really mixed...It's hard, my host community is kind of abstract almost, you know?
Focus Group 3	One of the things I've been battling—and I think a lot of us have...is apathy. So, to try to organize people, get together, to have some conversations about why are we coming together, what do you want to see happen, what can we do, and to know that from the very ground level in a distressed county, that folks can come together and support each other and connect each other to resources to help them do whatever it is...it's creating the social connections with each other and establishing that so that people can hear each other out, versus the division that is present.	Yeah, I feel that too...we're battling apathy. And also, being in a constant tangle with trust issues from the community, just because it's very divided and the people that are trying to do this work...are very separated from the people that actually need to benefit from that work that they're supposed to be doing. And trying to merge those two communities has been difficult.

collections that either did or didn't capture certain aspects of their lives. In the third focus group, the photos were barely even mentioned though they were in front of us. That final focus group, the culmination of a year of work, revolved around overarching perceptions of community organizing work in rural Appalachian communities. Through their exchanges, some about uncomfortable realities about power and privilege in their communities, the trust in the room became apparent.

While trust does improve research outcomes, leading to more in-depth findings and perhaps more nuanced discussions about problems, trust generated during a project can have outcomes that linger in communities long after a project concludes. Methods like PGI that provide opportunities for community members to spend time with one another and to talk about their experiences with clear prompts designed to bring them together strengthens trust between stakeholders. Productive, responsive solutions to public problems can emerge when people come together to share their experiences in ways that feel authentic, and visuals can support this goal.

Affordance Two: Capturing Individual and Collective Understandings of Abstract Concepts and Problems

Community-based research projects are often designed to collect perspectives from multiple points-of-view. Rather than depending on top-down approaches to defining problems, these projects tend to gather insight from community members who represent problems in different, sometimes contradictory ways. Projects that incorporate place-based visual methods provide participants an opportunity to document how abstract concepts, like inequality or injustice, shape their material lives. Creating visual representations that capture different aspects of their community allows participants to reflect upon their own perspectives, and sharing those representations provides opportunities for shared conversation on complex issues.

Participants in the pipeline development project consistently juxtaposed visual evidence of material changes in their communities alongside reflections on the admittedly abstract problems that caused those material changes. In this project, participants were asked to take pictures leading up to our go-along interviews. (In some cases, these were more of a drive-along experience in cars or even on ATVs than walk-alongs.) During those interviews, participants showed us around their property or larger community, telling us stories about how pipeline development had changed the landscape surrounding us. Naturally, participants would often pull up photos they had taken recently but also months or even years in the past.

At the most basic level, these place-based interviews and accompanying photos served as documentation for how participants' land and lives had changed over time. In many conversations, participants would start telling us about a certain event and then stop to say things like, "I have photos I'll send you of when they were doing it," (Participant 7B) or "I'm looking at some pictures now...okay, here's the picture" (Participant 14B). Participants showed us pictures of their homes, yards, and surrounding acreage, oftentimes organized into before pipeline construction and after-pipeline construction. (Over 60% of participants reported that they had taken photographs before we ever prompted them to, suggesting that pictures served as an important form of documentation during construction projects.)

During our conversations, however, these photographs often took on greater meaning as participants arranged them and wove them together through story. Pictures (or surroundings) became more than the literal image and instead conjured a much more complicated scenario, event, or feeling. In one interview, we were sitting on a front porch and looking out at the road leading up to a participants' property. They said:

> [As a] matter of fact, I was just looking at the pictures of [the mudslide]. It rained and washed down onto the highway there....There was actually two inches of mud on the highway past my mailbox almost halfway to you know, past my property. I'm like — there's nobody you can call because at the time...you called an 800 number, left a message. So a waste of my time, you know, trying to do that.
>
> *(Participant 13B)*

While the participant began to tell a story about a particular event, their examination of the pictures they had taken to merely document that event led them to a more substantive observation: the lack of communication channels between landowners and the gas company. While their inability to get in touch with a live person in a moment of crisis was frustrating, the gap between stakeholders was a significant tension for many participants in the project. For this interview, being present with the participant to observe the actual site of change, in addition to using photographs to conjure memories of what the land had looked like previously, led to more in-depth reflections that touched on systemic issues that seemed to surface for others.

During this project, I also learned how important photographs were for connecting with other community members who had faced similar issues. As noted above, many landowners took photographs regularly to document how development was impacting their property and community at large. Several participants noted that they shared these photographs publicly, through social media platforms, and at public meetings organized by citizen action groups to try and respond to company actions.

One participant shared that they had taken photos themselves of traffic concerns during development but had also reached out to others in the community to gather their photos to share at a public meeting. They said, "Everything...that was all brought out at these meetings. And citizens participated in that...I've got pictures...they gave me pictures of school buses meeting these trucks in blind turns. That went over big!" (32A). This particular collection of photos represented more than just school buses and trucks almost colliding with each other, but rather, serious safety issues in the community that pointed to, once again, issues regarding communication and collaboration between stakeholders.

Through visuals including live examination of their surroundings and photographs of prior experiences, participants deepened both their individual and collective understandings of how pipeline development had shaped their lives. By holding space to reflect on the greater significance of particular moments, participants were able to widen their perspectives and comment on how factors like power imbalances and lack of agency potentially put certain communities at risk during development. Visual place-based methods offer a way into these complicated conversations that are approachable and inviting in scenarios where direct questioning might not be.

Affordance Three: Promoting Visions of Ideal Futures

In each of the projects described in the previous section of this chapter, participants expressed feelings of gratitude for being in the same space as other people going through similar experiences. One artist in the collage project stated, "I feel so much better about my life being here with you all. I'm serious" (Listening Session 1, 2022). Moments of connection and even joy were common across the three studies as participants felt connected to others through the sharing of their stories through both narratives and visuals. These moments often turned to discussions about hopes for the future. In the collage project with artists, this was especially true, perhaps because participants created representations of their past, present, and future as artists in West Virginia. Even in listening sessions in which participants discussed somewhat weighty issues shaping life in their communities including addiction, poverty, discrimination, and isolation, they were eager to think about alternate, more promising futures in their communities.

Some conversations focused primarily on immediate changes that people in the room could facilitate. Many of our sessions involved discussions of specific projects, with participants suggesting changes that they could make to their own involvement. Participants even expressed a desire to continue having meetups like ours, where artists could just create and talk about what it's like being an artist without external pressures to create or network. One group actually asked if they could all meet up again for a "part two or something" (Listening Session 1, 2022), while a participant in another group explicitly described the future as one where "spaces that are available for people to gather and create art not for any other reason other than the feeling that can happen when you do that in community" (Listening Session 4, 2023). In sessions where participants didn't already know one another, contact information was oftentimes exchanged as participants imagined collaborative projects with these new community connections. These desires point to the importance of researcher commitment to the communities they work in. One way to build sustained engagement with community members is to partner with community organizations already working with community members and to design research that supports their goals. The collage project emerged from conversations with the arts organization I was working with, and so we were able to use their community connections and my expertise to build a project that promised outcomes beyond academic research.

Other articulations of the future were focused on more expansive shifts that involved a prioritization of art in their community—even if the people in the room that day were integrally involved in that shift. Our session became a space for strategizing as participants in several sessions listed out the ways that they were trying to build infrastructure so that art could be a central aspect of their communities, economically, culturally, and socially. One participant summarized their experience of our listening session as follows:

> [In this session] All I did was just envision a whole bunch of different people together as a community doing things together. And then I also listened. I was listening to the listening session instead of talking and I started pulling words that everyone was saying. So, other than we are all artists because that's what I'm envisioning, I've got: balanced, safe, creative, together, mountain folks, celebrated, empathetic, important, dreamers, Appalachian, fantastic, loved and loving, brave, joyful, accepted, free, kind, valuable, and connected.
>
> *(Listening Session 11, 2023)*

As participants created and shared their collages, talking about their overall experiences as artists in West Virginia, more immediate conversations about their communities and their hopes for those communities unfolded, demonstrating the value of creating visual texts in shared spaces with others.

For participants in this project, working with the visual medium of collage was a way to create space for the kinds of conversations in which people could dream. Rather than designing our listening sessions as traditional focus groups, by incorporating a visual and place-based activity, the space felt less formal and more generative. Using collage as a way into admittedly complex conversations about place prioritizes community perspectives and encourages participants to ponder actions outside of the immediate work of contributing to a research project.

Designing Visual Activities for Learning About Place-Based Problems

This chapter has argued that visual place-based methods are a useful tool as we try to address complicated problems that disproportionately affect marginalized communities; they are one of many strategies we can use to document and analyze the issues facing communities. Like other approaches that directly address the importance of lived experience (e.g., methods like ethnography or theoretical frameworks like intersectionality), visual place-based methods highlight the lived realities of environmental, economic, and social injustice that often go unnoticed.

Photographs, go-along interviews, and collage-making can be used in different ways, so what follows is just one possibility of many that could be used with a group of students, researchers, community members, and practitioners. These three research approaches could be used in conjunction with one another to create a **place-based problems portfolio** that documents multiple aspects of an issue. Rooting visual activities in place ensures that there is a shared environment and set of circumstances for participants which will ideally yield more shared observations and in turn, strategies for addressing problems. The activity as described below is organized so that a group can do this internally, but it could be adapted to bring in external stakeholders (i.e., in a classroom setting, students in the class could recruit students outside of the class; in a professional setting, practitioners could work with community members; and so on).

1 **Determine an issue of shared concern:** In your group, select an issue that affects a local community that you belong to. To do this effectively, you will need to define your community carefully (Is this community rooted in a particular locale? A global group that is less defined by geographic borders? A place-based community that exists in response to particular activities?)

2 **Research and write up findings:** Conduct some preliminary secondary research on the issue you have chosen to write a short research brief that outlines the background and factors contributing to this issue.

3 **Decide on parameters for inquiry:** After reviewing the research brief, consider what elements are under-represented or even missing from this account of the issue. What gaps seem to exist in your knowledge about the issue? Who are stakeholders who might be affected by the issue but whose perspectives seem to be missing? What further questions do you have about the issue? After generating a list of areas for development, consider whether more research is needed before moving on to the next steps.

4 **Conduct go-along interviews:** Amongst your group, arrange times to walk or ride along with other group members as they do an activity that is related to the issue you all have chosen. For example, if your issue is the building of a new power plant, and one group member plays soccer each week at the local park and feels that the power plant might impact air quality at the park, group members could attend a practice with that member. As you conduct your go-alongs, take notes on and even pictures of what you're learning.

5 **Generate prompts for inquiry:** Now that you've gotten a deeper understanding of the issue, come up with several questions that could prompt members of your group to take pictures that might depict how the issue of choice shapes their lives over a short period of time. It is best to have more open-ended questions so that you get a variety of photographs and so that everyone feels eager to interpret the prompts in their own ways.

6 **Take photographs:** Provide a week or so for taking photographs. Try to encourage one another to save or upload those photos into a shared folder regularly so that all of the data is easily accessible by one another.

7 **Reflect on photographs through writing or discussion:** Once photographs have been gathered, take time to reflect on photographs individually and collaboratively. For example, you might have people look at all their photographs and write brief reflections about what they see. You could ask people to look at everyone's photos on their own to identify trends, and then bring everyone together to talk about their photographs collaboratively, asking questions about similarities and differences between their experiences.

8 **Hold a collage-making session:** Now that you've gotten both place-based and visual representations of the issues' impact, it's time to reflect more expansively on those individual experiences and consider how they might relate to larger systemic issues. Get together with your group and create collages about how the issues make you feel, or what your

hopes or fears for attempting to address the issue hold. Try to provide a range of materials, including colored paper, magazines, flyers, stickers, and yes, photographs from earlier stages.

9 **Assemble portfolio:** Once all your primary research has been completed and you have taken time to look for patterns and trends across your visual data, you can assemble a portfolio of your findings. Using your earlier secondary research, the visual data, and potentially some additional research, you can put together an account of the issue that ideally amplifies previously unrecognized aspects of the issue and its impact on your community.

While the steps above can be adapted or re-arranged in a number of ways, at their core, each method of data collection is meant to allow for a range of perspectives on public problems that shape our lives. Using visual place-based methods as a way to learn more about our own experiences and those of others is a way that we can document the complicated realities of communities and train TPC students, researchers, and practitioners to be especially cognizant of how community members might perceive public problems.

Using visual methods in community-based research projects focused on place does have some drawbacks, including that such projects require a great deal of relationship building and organizational work. As a result, they are best suited for sustained coalition-building efforts that are designed to unfold over months or even years. Embracing methods like those outlined in this to piece prioritize relationship building can be part of a research approach rooted in slowness. Kuus (2015) argues that research engaging policy is "necessarily slow" and "must look at the broader social milieu" in which problems and solutions are based (p. 839).

Research methods like PGI, go-along interviews, and collage making are approachable and engaging ways to serve the goals of students, researchers, and practitioners alike. Such methods also serve goals outside of those immediate to research, providing opportunities to grow social relationships. Ultimately, visual place-based methods emphasize the value of community-based expertise and subsequently leverage that expertise to better understand the material implications of inequity and injustice—experiences that are often difficult to articulate in ways that reach new audiences.

References

Batova, T. (2021). "Picturing" xenophobia: Visual framing of masks during COVID-19 and its implications for advocacy in technical communication. *Journal of Business and Technical Communication, 35*(1), 50–56.

Carlson, E. B. (2020). Embracing a metic lens for community-based participatory research in technical communication. *Technical Communication Quarterly*, 29(4), 392–410.

Carlson, E. B., & Caretta, M. A. (2023). Collaborative sensemaking through photos: Using photovoice to study gas pipeline development in Appalachia *Qualitative Research*. https://doi.org/10.1177/14687941221149582

Carpiano, R. M. (2009). Come take a walk with me: The "go-along" interview as a novel method for studying the implications of place for health and well-being. *Health & Place*, 15(1), 263–272.

Chilton, G., & Scotti, V. (2014). Snipping, gluing, writing: The properties of collage as an arts-based research practice in art therapy. *Art Therapy: Journal of the American Art Therapy Association*, 31(4), 163–171.

Crenshaw, K. (1989). Demarginalizing the intersection of race and sex: A Black feminist critique of antidiscrimination doctrine, feminist theory and antiracist politics. *University of Chicago Legal Forum*, 1989, 139–168.

Gogan, B., & Scott, J. C. (2023). StoryMapping civic engagement: Reflexive chorography, spatial justice, and the carnegie classification for community engagement. *Communication Design Quarterly Review*, 11(4), 30–44.

Gonzales, L. (2022). *Designing multilingual experiences in technical communication*. University Press of Colorado.

Jones, N. N., & Walton, R. (2018). Using narratives to foster critical thinking about diversity and social justice. In A. M. Haas, & M. V. Eble (Eds.), *Key theoretical frameworks: Teaching technical communication in the twenty-first century* (pp. 241–267). University Press of Colorado.

Kuus, M. (2015). For slow research. *International Journal of Urban and Regional Research*, 39(4), 838–840.

Lippard, C. D., & Spann, M. G. (2014). Mexican immigrant experiences with discrimination in southern Appalachia. *Latino Studies*, 12, 374–398.

McNely, B. J. (2013). Visual research methods and communication design. In *Proceedings of the 31st ACM International Conference on Design of Communication* (pp. 123–132).

Nichols, G. W. (2013). The quiet country closet: Reconstructing a discourse for closeted rural experiences. *Present Tense*, 3(1). Retrieved from: https://www.presenttensejournal.org/volume-3/the-quiet-country-closet/

Reddy, N. (2019). "The spirit of our rural countryside": Toward an extracurricular pedagogy of place. *Community Literacy Journal*, 13(2), 69–87.

Rose, E. J., & Cardinal, A. (2021). Purpose and participation: Heuristics for planning, implementing, and reflecting on social justice work. *Equipping technical communicators for social justice work: Theories, methodologies, and pedagogies* (pp. 75–97). Utah State University Press.

Spriggs, B. L., & Paden, J. (2018). Introduction. In B. L. Spriggs, J. Paden, & F. X. Walker (Eds.), *Black bone* (pp. 17–19). University Press of Kentucky.

Swacha, K. Y. (2021). Living visual-voice as a community-based social justice research method in technical and professional communication. *Technical Communication Quarterly*, 30(4), 375–391.

Walton, R., Moore, K., & Jones, N. (2019). *Technical communication after the social justice turn: Building coalitions for action*. Routledge.

Walton, R., Zraly, M., & Mugengana, J. P. (2015). Values and validity: Navigating messiness in a community-based research project in Rwanda. *Technical Communication Quarterly*, 24(1), 45–69.

Wang, C., & Burris, M. A. (1997). Photovoice: Concept, methodology, and use for participatory needs assessment. *Health Education & Behavior*, 24(3), 369–387.

Weedon, J. S. (2017). Representation in engineering practice: A case study of framing in a student design group. *Technical Communication Quarterly, 26*(4), 361–378.

Wertz, O. M., Workman, K., & Carlson, E. B. (2023). Seeking out the stakeholders: Building coalitions to address cultural (in)equity through arts-based, community-engaged research. *Communication Design Quarterly, 11*(2), 18–27.

Young, I. M. (1990). *Justice and the politics of difference.* Princeton University Press.

4

EATING TO HEAL

Using Design Thinking to Reconceptualize a Community-Engaged Project for Health Justice

Sarah Moon

Introduction

Planning a community writing and performance project is something I learned "on the job," informed by scholarship and smart advisers like community writing scholars Tiffany Rousculp (2014) and Tom Deans (2000) The project I planned and carried out, Write Your Roots, served as the focal point for my dissertation. The origins of the project lay in my interest in public sphere theory. After reading Jurgen Habermas (2001), Michael Warner (2005), and Frank Farmer (2013), I became interested in how community writing could enter and influence the public sphere, specifically how performed personal narrative could serve to recontextualize the space in which public discourse on a given topic occurs. I chose food as the initial topic both because of its centrality to our lives, regardless of our demographic differences, and because of the progressive organizing efforts around food in our community. After two successful rounds of the project in which participants were asked to write about *any* aspect of their relationship to food (topics included teaching nutrition to school children, gardening, raising animals for food, working in a soup kitchen, and falling in love through cooking), for the third iteration of Write Your Roots, I decided to narrow the project focus to dietary changes made for medical reasons. I knew that given this more sensitive topic focus, I would need to plan more thoughtfully.

Reaching people who had had to change their diet for medical reasons and showing them what the project might offer them was not the same as appealing to the population at large to write about *any* aspect of their relationship to food. Given the challenge of targeting a much narrower

DOI: 10.4324/9781003469995-6

subset of the population and the sensitivity of the topic, recruitment, and planning would require greater knowledge and strategy on my part. Design thinking offered a much-needed structure for taking a more empathetic, informed approach to planning recruitment and revising the project structure.

This narrower topic focus introduced me to a subject that was new to me: community health. Positively, I found that there is a wealth of scholarship detailing the use and success of human-centered design to address community health concerns (Aflatoony et al., 2024; Solomon et al., 2023; Vechakul et al., 2015; Westgard et al., 2023). These studies collectively suggest a strong, natural link between design thinking and community health initiatives. While this third Write Your Roots is not a community health initiative per se, it is a health justice initiative that aims to raise awareness of and discussion around a significant health issue in a community context with the potential of helping community members in "secondary prevention," prevention with the purpose of reducing the impact of a disease or injury that has already occurred.

In this chapter, I will first provide an explanation of the original Write Your Roots project model. Next, I will provide a survey of the responses collected from interviews with seven individuals who had to change their diet for medical reasons. Finally, I will share my plans for the next Write Your Roots participant recruitment and project re-design based on these responses.

Inventing the Write Your Roots Project Model

To show the contrast of what design thinking brought to project planning for Write Your Roots, I will first summarize my planning process without design thinking for the original project. The inciting event for creating the first Write Your Roots was an email from the UConn Humanities Institute (UCHI) calling for proposals that made a contribution to facilitating public discourse. Around the same time, I learned about the then-new food-and-community-centered nonprofit organization in Willimantic, CT, CLiCK (Commercially Licensed Kitchen), which provides licensed kitchen space to small-scale food producers, nutrition, and cooking classes as well as low-cost ServSafe classes. My initial idea in responding to the UCHI call was a community writing center café at CLiCK.

However, after speaking with my dissertation adviser, I decided it would be wise to plan a smaller-scale project with a finite timeline, taking a more "tactical" approach. Advocating for a tactical approach to community writing, Paula Mathieu (2005) writes that "tactics seek rhetorically timely actions" and contrasts this with the long-term strategy that universities employ (p. 17). She cautions that "to apply strategic rules calls upon a potentially colonizing logic that seeks to control the space of the interaction through stability and long-term planning" (p. 17). Mathieu (2005) emphasizes that a tactical approach allows community writing project planners to more easily face "time challenges, incompatible schedules, the often conflicted spatial politics involved in deciding on whose turf work can and should take place" (p. 17). As the process developed, I found all of this to be true in the case of Write Your Roots and would add that a tactical approach allows a community writing project to enter into an already occurring process of community development and be shaped by dedicated community members' visions. So while it may be university scholars creating and facilitating the model, it is community members who set the community-at-large goals for the project. The tactical, community-driven model also helps to avoid the "sad breakup story" that seems to be common in the service learning partnerships between university faculty and non-profit organizations (Restaino & Cella, 2013, p. 253). Finally, the tactical approach makes it easier to re-imagine and re-design between individual project cycles.

I was attracted to food as a topic for community writing because of the way it serves as a basic common denominator for people of all different backgrounds. We all eat, and we all experience an ongoing struggle to balance our income, our cooking skills, our food access, our health, and our hunger. We all have food histories that help define us. In *The Practice of Everyday Life*, Michel de Certeau (1984) identifies cooking as one of the practices that afford opportunities to consumers who have been rendered "immigrants in a system too vast to be their own" (p. xx) to adapt "the dominant cultural economy...to their own interests and their own rules" (p. xiv). I was excited about the way that food can be an accessible channel for expression of self and culture, outside the oppressive forces of dominant culture.

Because of my background in theater and the collaboration with food-focused CLiCK, a friend suggested that for the project, I invite community members to write and then perform monologues about food. This model was inspired by my combined interests in playwriting, community theater, and public sphere theory. While not based on any existing model of theatrical community writing, my concept bore a similarity to the Community Conversations that took place at Pittsburgh's Community Literacy Center. The CLC's Community Conversations were the collaborative activity of "local teens, college mentors, and CLC staff," and focused on "addressing pressing community problems by means of oral and written intercultural communication, problem-solving strategies, and rhetorical performance" (Flower, 2008, p. 112). CLC researchers referred to their work as *literate social action*, a term I adopt to help describe the vision I had at the outset of planning to Write Your Roots. There is significant power to mindfully create a context for speech acts that aim to shift audience perception and spur action. As Charles Bazerman (2013) writes, "The face-to-face situation of spoken language, the physical environment and the visible social relations of co-participants often can lend a concreteness to rhetorical moments that help align participants to common understandings of what is going on and what is needed" (p. 71). With Write Your Roots, I sought to explore the possibilities of literate social action to draw attention to and even inspire active support for local social issues. While the CLC's community conversations were geared toward problem-solving, addressing issues such as a teen curfew, I was interested in how literate practice could help draw wider community attention to and support for *existing* proactive efforts, in this case around food.

To attract participants, I took a mass marketing approach. I worked to get the word out in as many places to as many people as possible. I went on a local radio station, ran a notice in the local newspaper, *The Willimantic Chronicle*, brought flyers to the local farmers market, and posted to social media. In the end, I recruited 10 people. We met for the first time in September 2016 at CLiCK. After meeting with writing coaches for 6 weeks, we held a potluck read-through of the finished drafts. Then, in April and May of 2017, we rehearsed the monologues, revising along the way, with blocking (movements on the stage), set and props. The project reached fruition in June 2017 with a performance at the town hall and a performance and potluck at CLiCK. I knew I would want to run the project again to test the model's replicability so carried out a second Write Your Roots in Providence, RI in 2020 (Moon, 2022). For the next two years, Write Your Roots was on hold while the pandemic ran its course. Then, in December 2022, I experienced a health crisis that changed my whole perspective on food and, in turn, the Write Your Roots project.

Rethinking the Write Your Roots Model

My reason for revisiting the Write Your Roots project concept was personal. I gave birth to my third child in June of 2022. Six weeks later, our family of five moved from Rhode Island back to Connecticut to be close to extended family. I returned to work remotely that October. As anyone could imagine, it was a stressful first six months postpartum. And my body kept the score. Just a couple of weeks before Christmas, I was hit with a bundle of symptoms that completely overwhelmed me. After a few visits with the doctor and a blood test, I had two diagnoses. One of my new conditions could be managed with medication but the other could not; it could only be managed through diet. The consequence of not managing it was pain. Suddenly, I could not consume coffee, chocolate, carbonated drinks, citrus, onion, hot spices, vinegar, cranberry, or sweets. My new condition completely changed my basic assumptions about how we, collectively, relate to food. I had gravitated toward food as a topic for community writing and public discourse because of my perception of it as a rich and wide medium for expression and a unifying element of our lives. In my mind, food was generally a source of pleasure and positive associations for people.

After I had to change my diet for health reasons, I realized how many people experience food as scary, isolating, and stressful without adequate support to navigate the experience—at least when first adapting to a new diet. Suddenly, the joy of food is dampened by (1) physical suffering caused by problem foods, (2) having to tell others your food restrictions, (3) having to guess or ask if a food contains something you can't eat, (4) not getting to eat foods you love, and 5) being suspected of "faking it" to get attention or, more mildly, simply not having your diet taken seriously.

I realized that I myself had not taken the experience of health-related dietary restrictions seriously in my prior conception of Write Your Roots. Adequate, professional advice and social support for dietary restrictions was a health justice issue I had failed to acknowledge in my basic assumptions about people's relationships to food. With this realization came both a sense of shame and a new curiosity. I wanted to connect with others who were going through the experience of medically driven diet restriction and learn from them. I wanted to understand how others handled the experience emotionally, socially, and practically. I wanted to compare how health providers had communicated with us and supported us, or not. I swiftly committed to myself that the next Write Your Roots would invite solely people who have had to change their diets for medical reasons.

My first testing of the waters for this new project focus came in August 2023 when I organized a one-day "Eating to Heal: Write Your Roots" workshop at a public library. Though some people expressed interest, when I arrived for the day of the workshop, excited to hear others' stories, no one showed up. I was disappointed and discouraged. I took a step back and meditated on the idea. In reflecting, I found that I truly believed that this focus was a meaningful one that could produce a beneficial experience for participants and help spark meaningful dialogue in an audience. I also thought the public discourse that could occur around this health justice issue had *greater* potential toward making a community impact than it had had with the more broadly conceived Write Your Roots.

However, to get to that culminating point of a community discussion around this topic, based on the lack of turnout for the one-day workshop, I recognized that I needed to first rethink how I would recruit participants. I had to recognize that I was asking people to take part in a project that requires exploring and sharing about a very private, possibly painful experience. I needed to plan thoughtfully and with sensitivity. To assist me in the process, I turned to design thinking.

The five stages of design thinking are Empathize—Define—Ideate—Prototype—Test (Dam, 2023). Framing the planning of this project in a design thinking context, the *problem* was recruiting committed participants who would follow through for the full duration of the process. The first step then was to understand and empathize with the type of people who would be participating, the "users" in design thinking terms. While my personal experience of having to change my diet helped with this step, it wasn't enough. I put the call out to a Windham (the larger town of which Willimantic is one part) Community Facebook page and heard back from one person. I decided that for this phase of the process, I didn't need to speak only with Willimantic residents, but could speak with anyone experiencing dietary restrictions. So I put the call out to my own personal social media accounts. I heard back from seven people and began conducting interviews right away. Because I was interested in qualitative data that would show in relief the diversity of personal experiences of this shared situation, I felt that spoken, in-person interviews would provide the richest and most detailed qualitative feedback.

My interviewees, who will be referred to by pseudonyms, were Simone, a Milwaukee-area healthcare professional in her mid-40s who proactively changed her diet after alarming cholesterol levels came back in multiple rounds of routine blood work. The second was Carrie, a Willimantic resident and former home aid worker who became vegan after she was diagnosed as diabetic. The next was Laura, a Seattle-based Starbucks executive in her mid-40s who was diagnosed with Celiac Disease in her early 20s. Then I spoke with Thomas, a Hartford, CT-based IT professional in his mid-50s who discovered an allergy to rennet—an ingredient in most cheeses, just a couple of years ago. Next came Rida, a Providence-based, Muslim paralegal in her late 20s who changed her diet after experiencing debilitating migraines. Then I interviewed Jane, a seamstress in her early 30s living in New Britain, CT who similarly had to change her diet after migraines and GI issues became frequent. Finally, I spoke with Nicole, a nutrition writer in her late 40s based in Riverside, CA who also began experiencing bad migraines as well as anxiety and depression that led to the most severe dietary restrictions of the group.

Just from looking at the demographics of the individuals who volunteered to speak with me, I gained valuable data. They were 80% female, all college-educated, and faced their dietary changes as adults. Six were white, one was Middle Eastern. This suggested to me who would be most comfortable participating in this Write Your Roots, with the caveat that I had been promoting the project through existing contacts of my own and just one public space. I filed away the fact that I would need to think about how to attract male participants in the project despite culturally established mores around men discussing health issues openly.

The questions that I asked the interviewees were as follows:

1 When did you find out you had to change your diet and how did you find out?
2 What was the process of changing like? How did you cope in the transition phase?
3 Do you feel like the story of having to change your diet for medical reasons is one you would find value in writing about in a group format?
4 Would you be comfortable reading a written piece about your experience for an audience?

In some cases, I added additional questions if led to by what the interviewee shared. The answers to the first two questions helped me to understand how others experience the process of finding out the need for dietary restrictions. For almost everyone, the process was not a straightforward one. In some cases (diabetes, high cholesterol, and celiac), tests can confirm that a person has the condition that will dictate a clear dietary change. In other cases (migraines, IBS, mental health, food allergy), the path to dietary change is not as clear-cut. It first must be discovered that overall diet is playing a role in the problem experienced. Sometimes, a doctor will assist, but often not. Those whose condition was not "confirmable" had been left to go through a process of self-guided research, elimination diets, out-of-pocket nutritionist consultations, and "chance" encounters that eventually led them to a place of relative clarity on the link between their diet and their health conditions. Thomas, for example, recounted how when his food allergy caused him to pass out on the street while walking a half block from a restaurant to his office, he ended up taken by ambulance to the emergency room. At the hospital, they got his blood pressure stabilized and sent him home without any explanation for what he had experienced or direction on what to do to prevent it from happening again.

One thing that came through in response to the first question was the fact that most primary care doctors are not trained in nutrition and thus do not advise in this area, or, if they do, advise in a very general manner. Simone noted that because she only had high cholesterol and not yet a condition like heart disease, her insurance would not pay for her to consult with a nutritionist:

> So then I was told, you should go vegan, you should cut these things, and then I asked, 'Can I meet with a dietician?' and was told, 'You already have to be sick to meet with one.' So you have to go find these people on your own. There are not a lot of resources to help people physically do this.

The lack of insurance-covered dietary guidance for medical issues linked to diet is one health justice issue that emerged from these interviews. If you are low income, you do not have the extra money to pay out of pocket to see a nutritionist. All you have is the health guidance of the doctors your insurance will pay for you to see. And this is often insufficient to help people properly adjust their diet to address their health issues and prevent them from getting worse.

The response to the second question varied widely. This was where I saw the greatest need in terms of my own increased understanding of people in the group I was targeting. I could not have extrapolated my experience onto those I spoke with. Laura shared that being diagnosed as celiac,

> It was very isolating. If you think about how much of your interaction and social life revolves around going out for dinner or breakfast. All of those things became like little war zones. Like can I eat this thing this friend made for me? And what utensils did she use to make it with? It was really tough.

Maggie commented that,

> By myself, it was fine. But if I'm around other people to celebrate, it's very challenging. That is a very challenging thing I've had to work through emotionally. I bring my food almost everywhere I go. That had to be a mindful plan. You have to heat it up beforehand. You need containers. And, oh, everyone having a pizza party! And then I feel odd - I almost had concerns of making other people feel bad.

However, some of the other respondents did not go through this type of experience. Simone took a very positive approach to her situation from the beginning.

> It was kind of like of a mind shift to me to realize, no, I *want* to do this. And that was a mental shift that made me feel more powerful and in

control. And so with my diet thing, I wanted to apply that attitude that I want to do this because I want to do better for my health. And when I explain it to people, I tell them I have chosen this rather than this was put upon me. My youngest son, my 12-year-old, had a million food allergies and he's grown out of most of them and even just explaining that, people would be just feeling sorry for him. And that was really wearing on me, so I kind of learned from that experience, so if I'm going to be communicating this to other people, I didn't want this to come across as "oh, poor you."

Rida also took a more positive attitude after a long journey of an often unhealthy relationship to food that had climaxed in orthorexia, a condition in which one is obsessed with eating healthy food. For her, dietary change became part of a spiritual shift in her life. She explained that she gave her dietary needs over to a higher power, praying and asking for guidance on her diet. In this way, she was able to settle into a diet that did not trigger migraines or GI issues. Thomas, as the person I interviewed with the most specific but also the most medically threatening food issue, felt accepting of his restriction, He shared, "I can't go out to Italian restaurants anymore. As much as I like pizza we make ourselves, it's still not the same." But he noted that he was only losing some favorite foods. His allergy doesn't affect, for example, Indian food. So unlike others who face restrictions of more pervasive ingredients, he still gets to enjoy the experience of eating out.

My findings helped me to better understand how different people experience dietary restrictions for medical reasons on a material level as well as emotionally and socially. I found that I couldn't assume that people would feel loss or negative emotions necessarily, though some do. I saw that planning and adapting over time were both important facets of adjusting positively to the change. I saw for some people the change had served as a catalyst for enlightenment in both spiritual and cultural spheres. While having a better general sense of others' experience of this issue, most critical to the planning of the project were the answers to the last two questions about Write Your Roots project participation.

Everybody except one person agreed that they would find value in writing about their story in a group context, either for personal benefit or for altruistic reasons. Nicole, a writer, whose dietary changes are extreme and relatively recent, for example, felt that it would be a welcome opportunity to explore her experience reflectively. She described being in "emergency"-mode with food for the past two years without the chance to step back and reflect how the changes she's made have affected different aspects of her life, especially her job writing recipes that she cannot now herself eat.

Maggie expressed both personal and altruistic reasons for being involved in the project:

> I would...really feel support by them and I would enjoy supporting them and hearing their struggles, their successes. I don't really know anybody that's...yeah, I've heard little murmurs about a diabetes support group but I've never found a time or a place.

Jane felt like it could help her on a practical level, "I guess so, in the sense of, if it were like meeting someone who had a similar issue and how you cope – how you dealt with it and how you trained yourself to say, okay, I can just avoid this thing."

Others, like Laura, Jane, Thomas, and Rida, expressed a primarily altruistic motivation for being involved, Thomas said that,

> I think I would, just in terms of maybe helping somebody else who might have a similar issue because I went so long not knowing what was causing it. I know I'm not the only person who has anaphylactic allergies and don't know what's causing it. I think any help I could give to someone else in a similar situation.

Simone expressed feeling like she was past the need for group exploration of her diet, especially because she had gone through many of the challenges years earlier when her young son was facing food allergies:

> I think, for me, I'm already managing it well. When I was managing a kid with these allergies, I was in a lot of groups and that was helpful then, but I don't feel the same need for that now.

What her experience did make me aware of was that it would be appropriate to invite, in addition to people dealing with dietary restrictions themselves, parents of children who have had to introduce dietary restrictions.

Moving to the final question, four of the people in the group said that they would be comfortable presenting their written piece in front of a public audience. Laura, for example, has a theater background, so her experience made her feel confident at the prospect of performing. Nicole, Rida, and Simone also expressed being comfortable with the idea of performing their work publicly. Maggie, Thomas, and Jane all expressed that they would prefer for somebody else to read their work. Their reasons included not being comfortable speaking in front of a large group and not being able to commit the time to rehearse and perform.

In the prior iterations of Write Your Roots, I had considered the authors performing their own pieces to be crucial to the project model because I felt it important that they continue to revise their work in light of rehearsing it and experiencing how words landed in performance. The fact that three of the seven interviewees said they would not want to perform their writing confirmed that I needed to rethink performing one's own writing as intrinsic to the project. Based on this response, for the next iteration of Write Your Roots, I will let authors know that at the end of the writing process, they will be invited to choose whether to perform their own pieces or have them performed by actors. This will require recruiting actors to the project, a new component. Given the nature of the work, I would want the actors to meet with the authors of the pieces they perform and for the authors to observe and provide notes at an early rehearsal to ensure the representation felt good to them. Given this new need, I will be seeking to partner with a university theater department or community theater.

Nicole Lariscy's work in Birmingham, Alabama provides an excellent model for this format for community writing and performance in which community members' stories are performed by others. She and her students, who form what they call the *HearTell Story Works*, have "worked with patients in an adult day center to tell their stories of mid-to-late stage dementia" and "patients at the local comprehensive AIDS/HIV center to tell their stories of stigma and survival" culminating in the play *One Clear Light* which was produced by a local Birmingham theater company (Lariscy, 2016, p. 129). Lariscy and her students also created an immersive theater play on health disparities in urban Birmingham neighborhood. For this production, Lariscy's students partnered with the Division of Preventive Medicine at their university and the United Way's Walking School Buses program with the goal of educating people about the systemic forces like redlining that led to what is now considered "unsafe" neighborhoods in Birmingham. One thing I like about the shift in Write Your Roots to having some stories told by actors is that it emphasizes that, like the projects created by *HearTell Story Works*, the goal of the project is to provide a platform for community engagement around a topic of importance, rather than simply to provide an expressive channel and support group for participants.

Ultimately, three important shifts have been made to the Write Your Roots project based on the understanding I gained through my interviews. First, I would be changing the format of the project to offer participants sharing their stories the choice of whether to perform or have an actor rehearse and perform the piece on their behalf. As a result, I would be partnering with a university theater department or community theater to provide actors and direction. This would then need to be one of the first steps in the planning process. I would be reaching out to Eastern Connecticut State University Theater Department, University of Connecticut Theater Department, and Windham Community Theater Guild with a partnership proposal.

For the recruiting of participants, I would stress both personal and altruistic benefits to participating and the option to perform. Toward benefits, I would explain that this group could provide support and suggestions for those still in the process of adapting to their new diet. I would also explain that sharing their stories in this format could serve to help others yet to or at the beginning of a journey of dietary restriction. Secondly, I would indicate that participants could choose whether to perform their stories or not. I would explain that the project would take place in two phases: monologue development and monologue rehearsal. Participants could opt-in to take part solely in the monologue development phase with majority remote meeting obligations and, at that point, pass their finished piece onto an actor, or they could take part in both phases with a majority in-person commitment for the rehearsal and performance phase. This would allow people to make a choice based both on their comfort level performing and the time and flexibility they have for meeting.

Finally, in putting out the call for participants, I will target individuals with medically-driven dietary restrictions or parents of children with restrictions by creating a flyer to be posted and shared at local hospitals, clinics, nutritionist, dietician, and pediatric offices. I also plan to partner with CLiCK and the Willimantic food co-op to disseminate the call to their listservs. I will again plan to promote the project on local cable access and radio but will center it on my own personal experience with dietary restriction to establish ethos and trust among those listening who might consider taking part. In doing so, I will be able to present the issue with an understanding that is also informed by the experience and perspective of those I interviewed.

Food is scary. I once may have scoffed at that assertion or associated it specifically with people who suffer from eating disorders. Now I understand that there are millions of people in this country for whom certain foods cause and exacerbate painful and sometimes life-threatening medical conditions. When speaking with Thomas about his anaphylactic allergy to rennet, he mentioned a young New York dancer named Orla Baxendale who had just died after eating a mislabeled cookie sold at Stew Leonard's grocery store (Moynihan, 2024). So, yes, food is scary: it can kill you. Yet most of us are bombarded with delicious images of decadent foods and marketing slogans encouraging us to indulge without question. Maggie Macha shared her cultural awakening of the degree to which we are misled that she experienced after her diabetes diagnosis, "I think the big box grocery stores, fast food, convenience stores really is what blew people's health right out the window. Dig a little under the surface, and it's plain as day." When you see your toddler holding a knife, you lunge to snatch it away from them because they don't know how it works and they could hurt themselves. In our culture, food is (with greater complexity, of course) the same way. We don't understand the dangers diet poses, often until it's too late. Taking a community health approach to secondary prevention in the realm of diet is a gift that Write Your Roots can offer if planned and orchestrated well. By applying design thinking to reconceptualizing Write Your Roots, I've gained a better sense of the health justice purpose this project might serve and how to recruit and work with its participants successfully and with compassion.

Design Activities and Learning Exercises

To adapt this application of design thinking to a service learning writing classroom, students could interview the targeted recipients of a community service initiative of the non-profit partner they're working with to explore whether they might be interested in a Write Your Roots-style project. Students could gather information about the relevant problem the interviewees are experiencing, how they would feel about writing about it in a group, and the potential benefits they would perceive in sharing their experience in a public format.

In a non-service learning course, students could select a health justice issue that intersects with their experience or the experience of a loved one, then research and write a summary of community health initiatives designed to address the issue. Then they could interview people they know who have had experience related to this issue to learn about their experience and their feelings about it. These interviews could be channeled toward suggested revisions of a community health initiative they learned about in their research.

References

Aflatoony, L., Hepburn, K., & Perkins, M. M. (2024). From empathy to action: Design thinking as a catalyst for community-based participatory research in dementia caregiving. *Design for Health*, 8(1), 24–45. https://doi.org/10.1080/24735132.2024.2307225

Bazerman, C. (2013). *A rhetoric of literate action: Literate action volume 1.* The WAC Clearinghouse; Parlor Press. https://wac.colostate.edu/books/perspectives/literateaction-v1/

Dam, R. F. (2023). The 5 stages in the design thinking process. Interaction Design Foundation. https://www.interaction-design.org/literature/article/5-stages-in-the-design-thinking-process

De Certeau, M. (1984). *The practice of everyday life*. Trans. Steven F. Rendall. University of California Press.

Deans, T. (2000). *Writing partnerships: Service-learning in composition*. National Council of Teachers of English.

Farmer, F. (2013). *After the public turn: Composition, counterpublics, and the citizen bricoleur*. Utah State University Press.

Flower, L. (2008). *Community literacy and the rhetoric of public engagement*. Southern Illinois University Press.

Habermas, J. (2001). *The structural transformation of the public sphere: An inquiry into a category of bourgeois society*. MIT Press.

Lariscy, N. (2016). Staging stories that heal: Boal and Freire in engaged composition. *Community Literacy Journal, 11*(1), 127–137.

Mathieu, P. (2005). *Tactics of hope: The public turn in English composition*. Boynton/Cook.

Moon, S. (2022). Write your roots disrupted: Community writing in performance in the time of COVID. *Community Literacy Journal, 16*(2), 121–131. Article 36. https://digitalcommons.fiu.edu/communityliteracy/vol16/iss2/36/

Moynihan, E. (2024, January 27). NYC dancer Orla Baxendale, who died after eating stew Leonard's cookie with peanuts, remembered as "radiant" performer. *New York Daily News*. https://www.nydailynews.com/2024/01/27/nyc-dancer-orla-baxendale-died-after-eating-stew-leonards-cookie-with-peanuts-remembered-as-talented-driven-performer/

Restaino, J., & Cella, L. (2013). *Unsustainable: Re-imagining community literacy, public writing, service-learning and the university*. Lexington Books.

Rousculp, T. (2014). *Rhetoric of respect: Recognizing change at a community writing center*. National Council of Teachers of English.

Solomon, E., Joa, B., Coffman, S., Faircloth, B., Altshuler, M., & Ku, B. (2023). Designing for community engagement: User-friendly refugee wellness center planning process and concept, a health design case study. *BMC Health Services Research, 23*(1), 1–10. https://doi.org/10.1186/s12913-023-10007-7

Vechakul, J., Shrimali, B. P., & Sandhu, J. S. (2015). Human-centered design as an approach for place-based innovation in public health: A case study from Oakland, California. *Maternal and Child Health Journal, 19*(12), 2552–2559. https://doi.org/10.1007/s10995-015-1787-x

Warner, M. (2005). *Publics and counterpublics*. Zone.

Westgard, C. M., Llatance, M. A., Calderón, L. F., Rojo, G. P., Young, M., & Orrego-Ferreyros, L. A. (2023). The creation of a field manual for community health workers to teach child health and development during home visits: A case study of participatory content creation. *Journal of Community Health, 48*(6), 975–981. https://doi.org/10.1007/s10900-023-01260-2

5

DESIGNING ETHICAL CONSTRAINTS TO ENABLE FLOURISHING IN AN ONLINE COMMUNITY

Steve Holmes and Jared Colton

Introduction

Design practitioners and students likely need no persuasion as to the role of constraint in an online community. Echoing Donald A. Norman's (2013) pioneering work on affordance, designers know or soon learn that forms constrain—that is, they limit and exclude certain actions and enable/include others. Yet, while designers are often used to the relationship between users and objects, tasks, or purposes as the sole critical relation to think about, the design justice shift in the field requires researchers, practitioners, and students to think about the form more broadly. Following Costanza-Chock (2020) and Tham (2022), design justice scholarship has articulated the need to consider what ideologies, thoughts, attitudes, and ethical habits are made possible or impossible through technological and design constraints.

While "community engaged" in technical and professional communication (TPC) scholarship often characterizes design or teaching practices aimed at service learning or activist work with cooperatives, non-profits, and progressive organizations, this chapter offers a supplement aimed at taking a step back and asking: how can specific ethical frameworks help to understand and promote positive interactions in line with social justice work in online communities? The term "community engaged"—and this is not necessarily a bad thing—often presupposes that any community a classroom, practitioner, or scholar is engaging with has ethical values worthy of supporting. Phrases like a "community of learning" carry positive connotations. However, as we've argued elsewhere (Walwema et al., 2022), it's still imperative to connect specific ethical frameworks to such

DOI: 10.4324/9781003469995-7

community-engaged, social justice work. Technical communicators may find some communities to be more ethical than others.

This chapter will continue this discussion by exploring the ethics of inclusivity in online community design, with a particular focus on social media content moderation that is activist oriented. While the social media moderators we interviewed work for different companies with different goals, we hope to demonstrate some of the lessons teachers and practitioners can learn from how activist-oriented practitioners moderate and design online communities with inclusivity in mind. We specifically hope to demonstrate how moderation in relationship to the ethics of online community design and maintenance is necessary for promoting social justice issues.

Social media content moderation (Frith, 2014) is a specific site at which design constraints meet ethical dilemmas. A number of researchers outside of TPC have explored how especially marginalized users create activist pushback against the ways in which online content moderation is exclusive of gender, race, and sexuality concerns (Coombes et al., 2022; Jackson, 2023; Leybold & Nadegger, 2023). In TPC, Potts and Trice (2022) have examined some of the generalized ways in which the dynamics of exclusion and inclusion occur in online community moderation. Jordan and Holmes (2024, forthcoming) have explored how flourishing and virtue ethics relate to individual content moderators' perspectives on ethical values in relationship to agency.

This chapter builds on this past work to examine the large-scale design choices made by large commercial entities—companies who serve user populations in the millions. Specifically, we asked how thinking about ethics relates to design justice. In what follows, we relay and analyze three practitioners' perspectives on specific ethical frameworks and values that are useful to promote flourishing for inclusivity purposes. While the insights we learned from these practitioners differed in many ways, one consistent theme that emerged across the interviews was something akin to what virtue ethics would call prudential judgment, or the ability to discern the character of others in a given moment and then make decisions based on utility in response to that discernment. Briefly, virtue ethics focuses on the semi-stable learned habits or dispositions that motivate flexible ethical behaviors. Unlike deontology (following the rules) or consequentialism (the ends justify the means), virtue ethics focuses on how individuals form habits of ethical conduct through the presence of structures and opportunities that allow them to act (or to not act) generously, patiently, or honestly. Over time, how we act becomes part of who we are and the design of digital communication systems to enable or discourage particular behaviors plays a larger role in how our ethical decision-making forms over time.

As our next section explains, we previously interviewed a number of industry practitioners who were engaged in social media and online gaming content moderation, and this led to interviewing other types of social media content moderators. We offer a loose qualitative analysis of three moderators' responses less with an eye toward generalizing the data and more toward offering anecdotal experience and analysis that shows promising ways to think about ethics, flourishing, and design elements as they relate to community-driven learning. After all, it is common for ethicists to analyze from the outside, per se, and after the fact of an unethical incident. In contrast to that practice, we wanted to hear from community designers themselves to get their take on ethics and social justice issues, and we hope that this chapter demonstrates the value of incorporating these voices into thinking through how community-based engagement in online spaces can learn from some of the ways that folks in the industry who manage millions of users have endeavored to address inclusivity.

Interview Data Collection

As part of a larger study, we hope to interview 70–100 practitioners related to content moderation projects under two IRBs (11730 & 2023-1016) (the latter with collaborator Rachael Jordan). Our research question for this larger project is: How do ethical frameworks inform and/or help explain individual and collaborative content moderation practices in online communities and social media platforms? Our selection criteria were to research lists of social media content moderators who work with large online populations that have sustained engagement, such as videogame communities. By large, we mean 1 million or more users because we are interested in exploring ethical frameworks related to DEI in the industry. We used LinkedIn and corporate email accounts to cold call participation and then used snowball sampling to receive recommendations for other participants to recruit. We initially conducted several interviews in the Fall of 2020, but we had to suspend the project due to the COVID-19 pandemic because of new administration responsibilities that took on a life of their own during the pandemic, as well as caregiving responsibilities for new family members and older children out of school respectively. We initiated interviews again in the Fall of 2023. Unfortunately, our recent participant data did not offer responses that would be suitable for inclusion in this edited collection.

However, three participants met our participant criteria and agreed to 30–45 minute interviews. These three interviews were conducted between November and December 2020. All signed consent forms and gave us explicit permission to include their names, companies, and quotes from the transcripts as part of future academic publications. All interviews occurred and were transcribed on Zoom, then edited for accuracy where needed.

Thus, in dialogue with editors Jialei Jiang and Jason Tham, we elected to feature more of a conventional ethical description analysis of three of our Fall 2020 interview subjects. To be clear, we are not holding up our analysis below as a full methods study representative of all social media moderators. Instead, we see our analysis of participant data more in the realm of interesting and suggestive anecdotes based on the actual testimony of participants who have been in key moderation roles that impact millions of users on a daily basis.

For IRB #11730 we asked a series of semi-structured but open-ended questions with follow-ups for clarification or elaboration where appropriate:

1 In your own words, please describe what you do (remember, that some readers may not be familiar with social media moderation).
2 At what point did you become interested in online community ethics? And why?
3 What ethical principles or ideas motivate how you approach your work? Could you provide some examples? Do you see these ideas reflected in the communities that you and your company support?
4 What type of positive ethical relationships or practices do you think online communities can help to create? What type of negative ethical relationships?
5 Describe a past or recent moment where you and or others in your organization have discussed or enacted a ban on a particular user behavior or communicative action. What were the ethical reasons given for either enacting or not enacting this ban?
6 If you haven't already touched on this answer, what are some of the most pressing ethical issues that you face in online community/social media moderation? Whether regarding your specific company/organization or the social media as a whole?
7 Is there anything else you'd like to add regarding social media moderation/online community ethics that we have not asked?

Below, we report on some of our findings, focusing on each interviewee. We then elaborate and extend some of the ethical frameworks that were implicitly or even explicitly suggested by our participants.

Moderation as Education in Roblox

Age and technology are complex design considerations. Elderly mobile phone users have different needs than children, preteens, adolescents, and adults between 18 and 70. As such, one of the most interesting interviews we conducted was with Laura Higgins, the then and current Senior Director of Community Safety & Civility at Roblox. Roblox is very popular, with an average of 43 million daily global users, and with close to 60% of its users being under the age of 13. Higgins's role effectively makes her a cross-unit leader or point person for any substantive discussions or design decisions at Roblox related to education and safety concerns. Unsurprisingly, Higgins indicated multiple times that Roblox takes content moderation very seriously given its young user base. Like any online platform, they have community standards that address national and international laws, such as child endangerment, but also informal policies such as threats of violence, bullying, and harassment. Higgins acknowledged that Roblox faces the same issues that any online platform will encounter: there are countless gray behavioral areas, such as Roblox's efforts to grapple with negative fights over the #AllLivesMatter hashtag that we will discuss below. Such user actions were not exactly bullying or harassment but were still behaviors that required moderation attention. Even the most well-thought-out conduct guidelines will likely never be able to always address the full range of emergent ethical dilemmas that users and contemporary circumstances pose.

As a case in point, Roblox's community standards discuss a commitment to inclusivity in identity and ethnic or cultural background in the Civility and Respect section. They state: "Roblox honors and welcomes users of all ages, backgrounds, and identities. You may not demean, threaten, or attack individuals or groups, or encourage others to do so directly or indirectly, on the basis of their gender, national origin, etc." (Roblox Community Standards, n.d., n.p.). The primary way that Roblox maintains this commitment on a daily basis is through automatic text chat filter analysis. According to the Community Standards, Roblox filters all text chat for "discriminatory speech, bullying, extremism, violence, sexual content, etc., as well as personal information and instructions on how to move off the platform" (n.p.). Higgins indicated that these filters are continuously updated especially when users start to creatively circumvent the filter by inserting a symbol like a $for an "S." A stricter filter set also exists for users under 13.

This text chat filter has been effective for maintaining many users' abilities to flourish in Roblox in the sense that they have the opportunity to form ethical habits by forming meaningful online friends and thereby continuing to have such opportunities. If users were subjected to constant harassment, then it stands to reason that their behaviors either shift to avoiding certain topics or forums, which means that users are depriving themselves of opportunities to form positive ethical habits. Higgins indicated, however, that not all censorship choices were as clear-cut as filtering curse words or common homophobic terms.

For example, while the activist organization and hashtag #BlackLives-Matter had been around since 2013, Higgins noted that the summer of 2019 had witnessed a noticeable spike in hostile text chat interactions surrounding the use of the words "AllLivesMatter." Even though users can select the groups and friends they chat with, enough negative interactions had occurred to impact player enjoyment, and the Civility team was asked to do something about it. Briefly, the terms #AllLivesMatter (and #BlueLivesMatter) were weaponized by the U.S. political right to create a straw person argument response to antiracist positions. Rather than engage the historical facts of Black oppression, structural racism, and disproportionate police violence against Black male bodies in the United States, many AllLivesMatter and BlueLivesMatter defenders tried to move the goalposts of the argument so to speak. They claimed that the BlackLivesMatter movement did not care about police safety or non-black deaths at the hands of police.

Thus, Higgins and her Civility team at Roblox were faced with an ethical problem: they needed to do something, but what? Should they consider banning disruptive players who used the term AllLivesMatter to combat or even troll progressive users? Higgins talked through her team's thought processes, which encompassed considerations such as whether it is even ethical in the first place to ban an 11-year-old who may not even be fully aware of the complex histories behind current political events. Also, Higgins wondered how Roblox could create an inclusive space where multiple political perspectives—including those who were not hospitable to antiracist positions—were welcomed while at the same time drawing a firmer line about trying to educate users about poorly conceived critiques of antiracism.

AllLivesMatter was causing enough problems that Higgins called an emergency meeting of multiple teams at Roblox to discuss that policy. She stated: "We decided rightfully that AllLivesMatter wouldn't be allowed on the platform. But we'd explain why. [addressed to an 11 year old year] 'We know why you might want to use this, but did you know that All-LivesMatter is being used to shut down Black voices?' For us it's about creating teachable moments, but it doesn't work everywhere." To be clear, Higgins and Roblox stopped short of banning users. In turn, users were welcome to use BlackLivesMatter and say critical things about antiracist attitudes or beliefs. What they could not do was invoke AllLivesMatter without triggering an educational design constraint.

Another reason not to ban had to do with maintaining user engagement with these educational opportunities. Higgins noted that she appealed to a sense of utility to her colleagues and superiors: if Roblox banned especially underaged users for protection, then users would switch to Discord or another chat app and be able to avoid all filter monitoring. Higgins indicated that she had spent time on the Discords that exist adjacent to Roblox. She commented that the younger users had grown up and noted that "because we built a platform that was so designed for kids…but kind of an age down [from teenagers]…now we've got these teenagers who are like 'this is so restrictive. This is not fun'" (19 min). On Discord, however, Higgins noted a prevalence of discussions around anxiety, depression, and self-harm (29 min), which meant that Roblox needed to determine how to moderate in a way that kept users on the platform instead of adjacent to it. She viewed this as an ethic of care, an obligation or duty, for Roblox's users, considering them even when they're not on the Roblox platform: "[Users are] still our community when they're on Discord, Twitter." Roblox clearly adheres to an ethic of protection in calling attention to the problems of BlueLivesMatters or AllLivesMatter. This ethic worked to protect forms of inclusivity from white racism while nevertheless trying to make it clear that other online communities likely couldn't model exactly what Roblox does and would need to consider their own sets of constraints in order to engage with social justice issues effectively.

iFixit and Discerning Utility

At the time of this interview Taylor Dixon was a Senior Technical Writer and Content Strategist at iFixit, a company of around 150+ employees that promotes, creates, and moderates do-it-yourself (DIY) guides and instructional manuals. iFixit also practices a social justice commitment to advocate for each user's "right to repair" the technology they own. This social justice initiative is based on the simple idea that "If you bought it, you should own it. Period." (Repair Association). In brief, all too often large companies will claim that a user cannot repair their own technology without voiding the warranty. The "right to repair" initiative is meant to empower users to know their property rights, including when such a statement about warranty voidance is not actually true and is just a scare tactic to encourage users to buy a new product.

In his role, Dixon had multiple responsibilities related to social media and online community moderation, including the following:

- Moderating the iFixit website, which is basically an open-source repair manual wiki site dedicated to accumulating and sharing repair knowledge for anyone who wants to fix any technology they own, on their own. These wikis are generated through crowdsourcing—users can upload instructions they've written to instruct others on how to repair a technology (such as mobile phones, turntables, video game equipment, and more).

- Producing content for iFixit, as well, including photos, video, blogs, and guides in order to generate traffic to the site—often taking a high-profile device from Apple or Sony to create a guide to repairing the device.
- Moderating iFixit social media, from their own question-and-answer forums and user-created guides to YouTube. Regarding YouTube, this included posting their own content on YouTube—e.g., refer to the video on taking apart a PS5 controller (link)—and moderating comments—often editing for accuracy and safety.

Dixon noted that he learned about why and how ethics matter to technical communication when he took an ethics and technology course in college, and that it helped prepare him for many of his work practices. He said that perhaps the most useful ethical framework for the work he did was utilitarianism (a term he remembered from the course) and that utility was analyzed and measured in multiple ways at iFixit, always weighing the benefits of publishing community-sourced content with the possible risks that content might pose to users.

Most importantly, he argued, the concerns were always in the context of making sure iFixit was useful to its users. "We are always analyzing content [from users and for users] in terms of its usability and utility: is this useful to site visitors and video viewers? Does it benefit [users] in a manner worth the time [that] moderating the content might take? If not, then it should be cut." But utility is not only concerned with usability, or at least not just a usability conceptualized in terms of ease of use. It's also interested in safety. Dixon mentioned that iFixit viewed risk in two ways: (1) Risk of personal safety and (2) risk to a person's device.

Dixon and other moderators would therefore cut, edit, or hold off on publishing content if the uploaded user content was potentially harmful to other users who might follow the content. In other words, user safety is an ethical concern that is "baked into everything we do," even the instructions for uploading instructions will have warnings: "Be careful how you describe a procedure." Some personal safety examples include if a user is repairing a laptop, could it potentially lead to a battery fire? If so, then it would be crucial that the DIY instructions made that clear and prominent; or, if such a risk was too high, then iFixit would not publish that content and notify the uploader with the reasons why.

According to Dixon, most users that interact with iFixit in some manner are aware of or have a general idea of usability and want to contribute something meaningful to the larger community. For example, they might not realize if taking apart a technology could cut another user's hand or if the instructions encouraged the user to do something illegal or that was unsafe to perform around others or without safety equipment. However, users who upload content don't always consider issues of safety at all, Dixon noted, or they assume that other users will intuitively know the risks.

Just as common is the risk to the person's device. In other words, Dixon hypothesized, is there a possibility that fixing one's own motherboard could break the motherboard? If a user follows instructions on the wiki or on YouTube, and as a result breaks their device, they could be upset that iFixit did not warn them of potential consequences. Anticipating such problems before publishing content, or making it clear to users that there are gaps in the instructions is crucial. In such cases, Dixon said that moderators would fix the content themselves if it was minor, notify the users, and give them time to fix it if it was more extensive edits, and if they felt there were still risks, they would include a warning to the users.

This focus on risk management extends past the wiki instructions but also to the health of the wiki as a whole and the community using it. Regarding the wiki, Dixon stated, "we moderate [the wiki] with the intent to create the greatest tool possible." Thus, community behavior also matters in this space. If users have a bad experience because of other users and don't feel like the moderators handled the situation correctly, they will be less likely to use iFixit and refer the site to friends.

Maintaining a thriving community is not easy (Howard, 2009). For iFixit, bots eliminate posts with profanity or bomb threats, and YouTube does the same. However, there are other ethical nuances, particularly with regard to the user experience of the wiki. As a result, iFixit has created a group they call "meta" where they discuss all iFixit-related issues. This is where "most of the company superstars hang out," according to Dixon, and they help guide on larger issues about moderation that might come up, such as if a user is violating a policy in a new way that might make the company need to reconsider the policy. It also is in these spaces that more discussions of the "right to repair," and whether the wiki is providing the necessary community support for this social justice initiative. Maintaining this community is crucial to enabling a robust political movement, so that users don't feel they are alone when a large company "uses scare tactics to make [them] think [they] can't repair, but this isn't illegal." However, even as these discussions can lead to or guide political advocacy, the site should also function for users who may not be interested in social justice: "it always comes back to utility."

Empathy Building Meets Utility in Two Hats' Moderation

Carlos Figueiredo is a snowball sample participant we received as a recommendation from another participant. Figueiredo joined the youth social media platform Club Penguin in 2008 as a full-time content moderator. At the time of our interview, Figueiredo had left Club Penguin to work at Two

Hat, which is a commercial content moderation company that designs anti-harassment solutions for a number of industry audiences. He is also active in the videogame industry having co-founded the international Fair Play Alliance in 2017. Similar to Higgins's response, Figueiredo viewed community design and moderation as a form of education over banning or punishing. Even for older audiences, Figueiredo indicated that platforms should "avoid the easy way out" as he referred to banning at one point.

While marginalized gender, ethnic, and sexual identities certainly take the disproportionate amount of toxic behavior in many online spaces, Figueiredo noted a positive for design thinking. According to him, the Two Hat gaming text chats were not overwhelmingly toxic. He stated that positive content was about "85% give or take [...] depending on the month or the week" on the level of user-generated text chat content that didn't run afoul of the automated text filter responses. Yet, he also conceded that as many as 27% of players for many popular online streaming games surveyed internally by his company indicated that they had changed their play styles in response to harassment or toxic player behavior.

In terms of design solutions for online community-based learning, Figueiredo stressed the need for companies to design ethical systems for users: "We're only as successful as we can give back to the community," thereby confirmed that good-intentioned individual community moderators likely would not be as effective as community's concerted commitment to ethical conduct. Tellingly, he stated "Ethics starts inside the company. Before the company is even formed." In other words, moderation to a certain extent in Figueiredo's experience was constrained by a company's internal commitment to particular ethical values.

Some of this ethical community design can work effectively through gamification. Many online forums are implicitly gamified by "likes" or points systems by users who violate the terms of service or community guidelines. Figueiredo noted an example of the game *Sky: Children of the Light* (2022) where social interaction is unlocked throughout the game. The game itself is aimed at younger users where cooperative and compassionate gameplay are required to advance in the game and to overcome certain in-game obstacles. By the 25-minute mark of that game, players can unlock more interactions by design, which procedurally means that the designers don't want players to be able to take social interaction for granted. It has to be earned and designers can foster it.

Figueiredo also raised a number of possible proactive design considerations to support inclusivity, such as changing the voice default settings (which are often male) for online multiplayer games when one wishes to use a voice filter instead of one's actual voice. He questioned how different inclusivity might feel if identity or gender were taken into account for random pick-up group selections. Rather than a reactive ban strategy, Figueiredo continually advocated for relating design and justice in terms of proactive work: what more can designers do to gamify or procedurally incentivize ethical incentives that also enable reaching a broader audience?

Despite the mostly benign posting percentages he mentioned, Figueiredo also acknowledged that 27% of gamers the Fair Play Alliance surveyed at one point indicated that they change how they play because of actual or potential harassment. He called for a total top-to-bottom design support of inclusivity: "How do we support the whole player journey from the moment they register? Communication systems are impoverished and it can be really difficult to 'get empathy,' if the system isn't built to help players find similarity with other players." Figueiredo argues that gaming communities need design that "humanizes" the experience where participants really see that there is another human they're interacting with: the strengths they have - that we have." Otherwise, players just use other players as a means to an end.

This is a clear design challenge raised by Figueiredo: designing with utilitarian concerns and empathy building in mind. If 27% of players are changing the way they play, then, in Figueiredo's words, "there's a moral responsibility to look at how we're designing games and social systems to remove as many blockers as we possibly can." He continued, "How do we design games so people connect as human beings" beyond just screaming at performance failures? Cooperation as a fundamental design activity needs to be more meaningful, substantive, and reflective. Echoing the Higgins interview, Figueiredo indicated that moderation and platform design should be more proactive about rewarding and incentivizing opportunities to form ethical habits rather than reactive to single acts.

One example he mentioned was an example of how Australian high school esports club teams for *League of Legends* (LoL) tried to incentivize better behavior. This idea was the brainchild of Kimberly Fall, who worked at Riot Games. Figueiredo also participated in this initiative by creating instructional design resources about online etiquette and online behavior regarding how to self-reflect before, during, and after an LoL club Match. Riot has over 100 million players a year in its online spaces. Players are supposed to follow the Summoner's Code, which establishes rules for respectful interactions. Figueiredo helped Fall and others at Riot Games establish a supplemental series of guides for LoL high school clubs. He called this step a "Seed of how gaming companies" can promote structures conducive to forming positive communication habits. Echoing Laura's point of view on Roblox's ethical responsibility, Figueiredo argued that "players going to go on...be active members [of other platforms and online communities] later [e.g., after they leave the moderation space]" so it is critical to build the foundational mentality of not just designing for "my game" or "my community." Instead, Figueiredo argued for designing and training players with an eye toward how they would behave in future systems that Two Hat or an original platform designer would not control.

Discussion of Ethics and Social Justice

What emerged from our interviews was the idea that there isn't a single ethical principle or even framework that is appropriate for all-inclusive actions in social media moderation. However, there was clearly a mix of utilitarian, care, and virtue ethics thinking expressed by each interviewee; regarding the latter, perhaps the most consistent virtue tacitly discussed would be something like prudential judgment, a virtue dedicated to the discernment of others' actions. As Higgins's interview confirmed, how an 11-year-old user is treated as more of a singular being in need of mentoring or as Figueiredo's interview indicated with regard to how mentorship for LoL clubs is required for high school players, how a user should be treated will depend on situatedness and the values that a community holds. For risk management in iFixit, one user might need significant feedback to revise their uploaded instructions, others might require a minor edit that the moderator could address on their own. These concerns also grow. Higgins noted how Roblox balanced education versus banning changed as the platform's early generation of users grew up from pre-teens to teenagers who pushed back more on explicit efforts to moderate behavior.

Thus, to reiterate, the "meta-"ethics we see emerging from these community designers is not a single framework like utilitarianism or deontology. Instead, it's a commitment to prudential judgment, or the ability to discern right or wrong actions in contingent circumstances. Aristotle included this as part of *phronesis*, also called practical wisdom, but other non-Western traditions have terms for it as well, such as the Confucian term *zhi*. Echoing previous work in TPC (Dragga, 1999; Wang & Gu, 2022), Confucian virtue ethics frames *zhi* as a disposition, a habit, related to one's character, which comes about through repetition and ritual. Once exemplars (notable examples of virtuous conduct) or broader virtuous patterns are recognized by others, *zhi* becomes known as wisdom. Van Norden (2007) summarizes the Confucian version of this virtue as

(1) the disposition to properly evaluate the characters of others and oneself; (2) skill at means-ends deliberation: the ability to deliberate well about the best means to achieve given ends, and to determine the likely consequences of various courses of action; (3) an appreciation of and commitment to virtuous behavior; and (4) intellectual understanding [...]."

(as cited in Vallor, 2016, pp. 106–107)

The concept of *zhi* can be useful and important as a lens for ethics and social justice in online communities because it describes an ethical disposition for how to apply virtue and utilitarian approaches. Rather than pit virtue ethics against utilitarian thinking, for example, *zhi* frames prudential judgment as a requirement for thinking about utility in an ethical way. *Zhi* assumes that means-end calculations are going to be part of how ethical decisions are made by groups, whether the focus is on families or larger communities. Social justice scholars are perhaps more familiar with the critique of neoliberal utilitarianism, which is a common theme in theoretical work that is suspicious of capitalism. In neoliberal utilitarianism, the majority seems to always be privileged over minoritized groups, and capitalism's ideology of radical deregulation and free market individualism is held to be the epitome of helping channel the major's individual decisions into policy and cultural practices that support the majority (Harvey, 2007). While this critique is important, what is less common are explicit acknowledgments that activists regularly rely on different forms of utilitarian thinking as a supplementary framework to enact social justice and pursue inclusivity. Thus, knowing when to apply utility as the right thing to do is situational, whether conceptualizing an ethical decision in terms of the greater good, risk management, pleasure/pain, or certainty/uncertainty.

In this way, Confucius's efforts to theorize *zhi* require that utilitarian concerns exist in the plural. Across divergent ethical situations, there will be many different ways to apply limited or expansive forms of utilitarianism, many of which will likely never be a commitment to utilitarian thinking in totality. There may be some sort of "first principle" like care or concern for children's well-being as a way to limit or prevent some of the problems that utilitarianism is prone to, such as inflicting harm on a minority if this harm creates the most amount of good for the majority. Another important consideration is that the use of any ethical framework can be situational and temporal. Roblox's decision to respond to #AllLivesMatter in the way that they did was a short-term *kairotic* response. Higgins confirmed that this response to this particular political circumstance would not need to continue forever as users' attitudes or broader global political situations change. Similarly, iFixit must discern when to balance community management concerns to its users versus when to ask their users to join the right-to-repair initiative by writing to legislators. In other words, iFixit seeks to maintain a thriving community based around the idea of DIY, while also pursuing social justice goals. These ends, while not mutually exclusive, can sometimes appear to work at odds. To invoke a more conventional technical communication distinction, there are appeals that are more functional versus appeals that are more expressive or persuasive in terms of aspirations of social justice. Importantly, Dixon's interview further highlights the need to avoid an either-or approach to thinking about utilitarianism. Not all utilitarian remedies are inherently good or bad, nor are they fixed and unchangeable. Context, circumstance, situatedness, and positionality all matter as do the ultimate ways in which pleasure, pain, and the "majority" of users' happiness are understood and articulated.

Another important implication of our data is that moderation cannot satisfy all singular needs at all times—another reason prudential judgment emerges as a crucial ethics consideration. Ethicist Shannon Vallor confirms that prudence means flexible thinking in relationship to what a moderator or a group of moderators in a company or organization have the ability to impact:

> To be right, the end must actually be achievable by the person in that situation, and likely to promote the ultimate aim of all human action: human flourishing or living well (which for social animals like us, always means living well with others). A virtuous person reliably discerns and employs effective practical means to achieve these ends, as appropriate to the specific circumstances in which [they] find [themselves].
>
> *(p. 37)*

Designing ethics through prudential judgment requires that moderation decisions have to be tweaked according to the prevailing circumstances. It would not be prudent to ban an 11 years old while it certainly might be to ban a 40 years old for the same behavior. Education versus banning as moderation will require different enactments in different platforms and communities. This decision makes sense because a lot of times moderation is almost figured as reactive. Users make decisions that are harmful or toxic to other users and then moderators need to react and curb those efforts. As such, banning often seems like the easiest tool in the moderator's tool kit to use, such as when Ravelry, the large knitting and crocheting social media community, banned support for Trump as support for white supremacy (Basu, 2020).

Yet, moderation is just one part of a diverse and complex platform assemblage for an online community. Prudential judgment in design connotes the ability to produce structures that shape civility and that starts at the top of a company, as Higgin's interview confirmed. Civility was the central "pillar of the company" at Roblox as Higgins put it. As such, she and other employees at Roblox have a wider spectrum of tools and support to try to make design decisions to ensure the flourishing of as many users as possible. Each interviewee tacitly made the argument that having an entire company or organization devoted to a particular set of inclusive values is more effective than any amount of "white hat" moderation by a single motivated moderator in a large online community. Prudential judgment in moderation is part of platform design at the end of the day in the sense that the platform and its values set the utilitarian conditions under which moderation will be enacted and exercised.

As such, platform designers can be more proactive in offering structural ways to support moderators' ability to enact prudential reasoning. Figueiredo noted that instructive parallel with his examples of video games that made players progress in particular behaviors before unlocking social interactions. A likely reason why trolling and harassment occur in online spaces is because it is too easy. The approach should not be to wait for trolls and then to respond, but instead to create inclusive structures that disincentivize trolling and promote respectful and healthy debate and deliberation to support users from a broad range of political perspectives. Even if bans that align with our social justice values might seem like ethically worthy remedies, the reality—as our interviewees have confirmed—is that banning should likely be a tool of absolute last resort and could even undercut the social justice initiatives of that community.

Conclusion

We are struck in the end by how the conditions to exercise prudential judgment should be more explicitly embedded within design contexts to support community flourishing. Prudence here does not mean that designers or moderators need to trust that players will somehow just magically "do the right thing" in terms of creating ethical habits. Indeed, anonymity online can give protections to marginalized identities' freedom of community in the same stroke that it can enable trolls to harass marginalized users and content creators. This is likely a harder lesson for community-based design thinking. When Facebook groups or Reddit forums are launched because they're streamlined and easy, we have to accept the sort of broad utilitarian constraints where a lot of awful harassment or hate speech is tolerated as long as it does not violate the Terms of Service or US/international laws. Yet, moderators (and users) can choose to support companies or subforums on platforms that do take responsibility seriously. As Figueiredo's example of the LoL high school clubs demonstrates, we can use technical communication and instructional design to intervene where moderation is not possible or cared about sufficiently.

Moderation exists on an assemblage and can benefit from when companies promote protective educational spaces adjacent to or within larger online communities to educate and guide rather than just throwing up one's hands and saying "I can't make Facebook change their minds about what behaviors they will and won't tolerate." Here, paradoxically, because Facebook is a diverse assemblage, sometimes critical thinkers hit something of an epistemic blockage because they feel frustrated that we will not be able to change the entirety of an assemblage. At the same time, better understanding the actual sites of intervention within a complex assemblage can provide concrete and actionable areas to intervene, even if the result might be on a smaller scale than we had hoped. As our analysis suggests, small differences scaled up to millions of users have great power to shape users' habits in online communities in powerful ways.

Learning Exercise

1 Use a university library database to locate and read Shannon Vallor's article "Social Media and the Virtues" about how ethical communication has changed in a networked age. Then, in groups, consider defining your own virtues and vices in relation to social media communication ethics. Which virtues should moderators use generally? Which ones should they use if inclusivity is an aim?

2 Study the Terms of Service and community guidelines for a specific online community or subforum for inclusivity purposes. Determine which utilitarianism principles exist. Pay special attention to which communication behaviors are proactively encouraged with utilitarianism frameworks. Do the latter relate to support inclusivity purposes? If so, how? If not, then try to think of ways that they could be revised to do so.

3 Search online for news stories in the past 3 years that deal with banning issues related to diversity, equity and inclusion. Read Confucius' relevant passages on *zhi* or prudence from *The Analects* that your instructor has provided and determine if the actions described in your news article demonstrate prudential judgment or challenge it. If the actions were not prudent, be sure to identify 2–3 new design decisions that could improve the opportunities for the actors involved to act with better judgment.

4 Search philosophy journals through your library database for the keyword "utilitarianism." Try to locate 2–3 different variations of utilitarianism. Identify points of overlap and departure in terms of how you might apply either to impact moderation behavior or online community values in the social media platform of your choice.

References

Basu, T. (2020). How a ban on pro-trump knitters unraveled the online knitting world. *MIT Technology Review* https://www.technologyreview.com/2020/03/06/905472/ravelry-ban-on-pro-trump-patterns-unraveled-the-online-knitting-world-censorship-free/

Coombes, E., Wolf, A., Blunt, D., & Sparks, K. (2022). Disabled sex workers' fight for digital rights, platform accessibility, and design justice. *Disability Studies Quarterly, 42*(2). https://dsq-sds.org/index.php/dsq/article/view/9097

Costanza-Chock, S. (2020). *Design justice: Community-led practices to build the worlds we need.* MIT Press.

Dragga, S. (1999). Ethical intercultural technical communication: Looking through the lens of confucian ethics. *Technical Communication Quarterly, 8*(4), 365–381.

Frith, J. (2014). Forum moderation as technical communication: The social web and employment opportunities for technical communicators. *Technical Communication, 61*(3), 173–184.

Harvey, D. (2007). *A brief history of neoliberalism.* Oxford University Press.

Howard, T. (2009). *Design to thrive: Creating social networks and online communities that last.* Morgan Kaufmann.

Jackson, D. (2023). Content takedowns and activist organizing: Impact of social media content moderation on activists and organizing. *Policy and Internet, 15*(4), 498–511.

Jordan, R., & Holmes, S. (2024, forthcoming). Inclusive flourishing as activist content moderation practice. In E. Frost et al. (Eds.). *Practicing digital activisms.* The WAC Clearinghouse.

Leybold, M., & Nadegger, M. (2023). Overcoming communicative separation for stigma reconstruction: How pole dancers fight content moderation on Instagram. *Organization*, Online first. https://doi.org/10.1177/13505084221145635

Norman, D. (2013). *The design of everyday things: Revised and expanded edition.* Basic Books.

Potts, L., & Trice, M. (2022). Digital community moderation values: Politics, news, and hot beverages on Reddit. *Proceedings of 2022 IEEE International Professional Communication Conference (ProComm), Limerick, Ireland* (pp. 133–139).

Repair Association. (2024). Our mission and history. https://www.repair.org/stand-up

Roblox Community Standards (n.d.). https://en.help.roblox.com/hc/en-us/articles/203313410-Roblox-Community-Standards

Tham, J. (2022). Pasts and futures of design thinking: Implications for technical communication. *IEEE Transactions on Professional Communication, 65*(2), 261–279.

Vallor, S. (2016). *Technology and the virtues: A philosophical guide to a future worth wanting.* Oxford University Press.

Van Norden, B. W. (2007). *Virtue ethics and consequentialism in early Chinese philosophy.* Cambridge University.

Walwema, J., Colton, J. S., & Holmes, S. (2022). Introduction to special issue on 21st-century ethics in technical communication: Ethics and the social justice movement in technical and professional communication. *Journal of Business and Technical Communication, 36*(3), 257–269.

Wang, X., & Gu, B. (2022). Ethical dimensions of app designs: A case study of photo-and video-editing apps. *Journal of Business and Technical Communication, 36*(3), 355–400.

6

INCORPORATING COMMUNITY KNOWLEDGE IN DESIGN

A Reflective Account of Designing Technology with Justice

Sweta Baniya, Katrina Powell, Ahoo Salem, Layla Scott, and Margaret Webb

Introduction

When migrant and refugee communities resettle in the United States, they encounter significant challenges related to housing, employment, education, healthcare, and cultural and technological adaptation. A crucial aspect of these challenges is the need to navigate complex federal and state policies written in English, which are often not tailored to individuals for whom English is a second language (McCaffrey & Taha, 2019). Moreover, policies and the accompanying resources intended to facilitate compliance with these policies require a high level of digital literacy, often facilitated by service providers and volunteers working with community members. These policies are often based on assumptions that migrants constitute a homogeneous group, that resources are uniformly distributed across rural, suburban, and urban communities, and that information technologies available to serve migrants are equally accessible (Alam & Imran, 2015; Bletscher, 2020; Hester et al., 2021).

Our research underscores the challenges faced by new Americans pursuing citizenship, who must navigate intricate federal and state policies that are not tailored to individuals for whom English is a second language. These challenges place undue pressure on migrants, including those seeking citizenship in the US. To attain citizenship, individuals must pass a test administered in English and respond to short essay questions verbally. Test takers are expected to demonstrate knowledge of US history, geography, and politics, despite potentially lacking formal education or proficiency in English. The cultural complexities, linguistic diversity, and socioeconomic factors

DOI: 10.4324/9781003469995-8

affecting these individuals influence their integration and well-being. Additionally, technological advancements create further challenges and barriers to how migrants access, interact with, and disseminate information. Therefore, our work aims to address these issues by leveraging knowledge justice theory to develop a website and mobile application tailored to the complexities of attaining citizenship.

Engaging refugee and migrant communities in the decision-making process and design of such applications facilitates what we refer to as "epistemic integration," an approach grounded in knowledge justice that seeks to integrate diverse sources of knowledge (Allen, 2018). We demonstrate how various technological tools can be designed with an emphasis on epistemic integration, highlighting their potential adoption by technical communication scholars and practitioners. In this chapter, we redefine ethical engagement in community-engaged scholarship, teaching, and learning, drawing on our pedagogical experiences and years of community-engaged work in diverse local and global communities. Given the rapid global changes and challenges, we argue for radical shifts in our approaches to community engagement, such that our actions as scholars contribute meaningfully to the communities where we live, work, and thrive. We reflect on the notion of reciprocity, suggesting that it sometimes lacks balance due to differing goals between academic and community members.

Our chapter begins with an overview of our history of community partnerships, followed by sections detailing our community partner and their background and reflection. We also include reflections from two students (undergraduate and graduate) and conclude with reflections from two faculty members on our work supporting refugee and immigrant communities, as well as our methodological choices emphasizing social justice in design.

History and Community Partnership Overview

This research project (IRB approved, #17-902) began several years ago when Powell and another colleague worked with local refugee and immigrant neighbors to understand the barriers and assets to integration they faced in the community. As research in refugee/immigration shows, "integration" is a loaded term, and the US government's conception of integration has very specific markers for successful integration into US society, including "self-sufficiency" (itself a loaded term), employment, and education.

In a 2017 pilot study with service providers and community members (including people involved with Blue Ridge Literacy, our project study site, and a local literacy center providing citizenship classes to immigrants in Southwest Virginia), we conducted ethnographic interviews and concluded that current information and communication technologies (ICTs) are creating more barriers than ever and not supporting meaningful integration (Pour-chot et al., 2018; Powell et al., 2019). Powell's three-year pilot study aimed to learn more about how ICTs reflect community concerns. Our participants reported that the ICTs available to them are limited in their ability to communicate information about extant resources which in turn can increase discrepancies in the ways those resources are distributed. However, these socio-technological networks, and the knowledge created by them, are (un-intentionally) hidden, not formalized, and difficult to access and navigate for refugees. We recognized that migrants/refugees and service providers were equally hindered by the same problem: a lack of ICTs with access to relevant, timely information informed by their knowledge and needs. Through our work with our collaborators, we came to understand that these concepts, and the policies driving integration efforts (like the requirement for refugees to be employed 90 days upon entering the country), did not align with new-comer experience, did not account for resources available in rural locations, and did not offer technological assets that accounted for newcomer digital literacies. After interviewing people in our local community, we argued for what Powell and Hester (2018) call "epistemic integration,"—a way of developing technologies for integration efforts that are designed using the knowledge and assets of newcomers themselves.

The results of the pilot study, together with curricular work incorporating outreach projects into classes addressing technical communication and tangible digital outcomes, led to the American Council for Learned Societies (ACLS) grant sponsoring the next phase of the work. Led by Sweta Baniya, our team engaged in co-designed technology, where students, scholars, and community members worked together to imagine technologies that were the most beneficial to them. Our community partners requested that we help them develop accessible technologies to address these challenges. Together then, the team developed a digital justice approach to ICTs to provide much-needed equitable access for community members seeking citizenship. With funding from the ACLS, Baniya and Powell, together with an interdisciplinary team, conducted focus groups, ICT development, user testing, and training workshops, to develop a prototype application and web design taking community needs into account. Working with community members via a participatory action-based research and design approach, we reimagined digital justice for refugee/migrant communities. Our asset-based approach allows communities to participate in sharing their knowledge and insights to design a mobile-based application that supports them in obtaining citizenship.

Working collaboratively with the community, our project ensured that all community members had equal access to technology and media as both producers and consumers (Bletscher, 2020). Since our mobile application is specifically targeting citizenship, our goal was to provide the informational infrastructure that not only creates greater information access to members of the community but also ensures the co-creation of knowledge (Fedyuk & Zentai, 2018). Additionally, the supplementary website adjacent to the mobile application provides supplemental information. We have co-created this digital resource with our participants, targeting citizenship processes and exams, by grounding these materials and resources in intercultural communication design to support communities with low literacy levels and lower levels of English languages (Baniya, 2022; Baniya et al., 2021, 2022; Jones, 2016). The website, together with materials specifically geared towards immigrant communities and their journey towards citizenship, will support communities typically underserved by ICTs.

Background on Blue Ridge Literacy Partnership & Reflection

The following section was written by Ahoo Salem, the Executive Director of Blue Ridge Literacy (BRL), a community-based literacy organization in Roanoke, Virginia. Salem's reflection highlights BRL's mission, services, and the pivotal role the organization plays in supporting refugee and immigrant communities in their pursuit of U.S. citizenship. Notably, Katy Powell and Sweta Baniya, the faculty members spearheading this digital justice project, have been collaborating with BRL for several years through previously established partnerships and existing grant-funded initiatives. This long-standing relationship allowed for a seamless integration of the current project with BRL's ongoing efforts, facilitating direct engagement with the learners and instructors involved in the citizenship preparation programs. Salem's contribution also underscores the symbiotic nature of this collaboration, where academic research aligns with the practical needs of the community, fostering mutual benefit and innovation. Established in 1985, Blue Ridge Literacy (BRL) is a Community-Based Literacy Organization (CBLO) located in the Main Library building in Downtown Roanoke, Virginia. The organization was established by two local librarians who saw the need for English literacy services among American-born adults in the Roanoke Valley. Over the past thirty-seven years, and per demographic changes in our community, Blue Ridge Literacy has adapted and expanded its services to address the English and civic literacy needs of our New American community. Blue Ridge Literacy offers one-on-one tutoring, leveled English for Speakers of Other Languages (ESOL) Classes, English and Digital Literacy services, and Citizenship Preparation programs.

The mission of Blue Ridge Literacy is to help adults achieve their life goals through literacy skills. The life goals of BRL's adult learners reflect the diversity of lived experiences, backgrounds, socio-economic status, and

English proficiency levels of our learners. We commonly see how learners adjust their literacy and life goals after achieving the initial skill they sought. For many lawful permanent residents, an important goal is mastering the English and Civics skills needed to become a Naturalized American Citizen.

As already alluded to, the current Naturalization test and interview consists of a civics section as well as an English proficiency test. While most learners focus on learning the civics test material, lack of English skills is often the reason behind failing a test. BRL's Citizenship Preparation Services are designed considering the learners' skill-based needs and naturalization processes. BRL offers beginner-level English and Civics classes for lawful permanent residents at the earlier stages of their immigration journey, as well as high intermediate/advanced level Citizenship Preparation classes and Citizenship Study groups for foreign-born adults who have started their naturalization process.

The application process to become a US citizen is lengthy and often complicated. In addition to the application costs, applicants need to budget for legal fees and commuting to USCIS field offices for their biometrics appointments, naturalization tests, and even oath ceremonies. Successfully navigating each step requires access to information and knowledge about support systems. BRL's Citizenship Preparation classes also serve as valuable peer experience-sharing resources. Learners who take their citizenship test share their experience with classmates and exchange tips and suggestions about the interview, transportation, and accommodation tips for visiting the USCIS offices in Charleston, West Virginia for biometrics appointments and Norfolk, VA, for their interview and oath ceremonies.

At Blue Ridge Literacy, we worked with our community partners (i.e., people seeking refuge within BRL along with literacy center volunteers supporting their preparation) to create spaces of connection and trust-building between our learners, as the recipients of services, and our community's service providers. Reflecting on our partnership with Dr. Baniya and her research team, the significance lies in how the academic work directly mirrors the needs of our learners. Aligning research efforts with the practical realities of our citizenship preparation programs ensures that the project's initiatives are tailored to address our diverse learner base's specific challenges and requirements. This approach enhances the relevance and effectiveness of the research team's proposed tool and underscores the importance of community-centered design. By involving the voices of the community in guiding and shaping their application, the research team ensured that firsthand experiences and perspectives of adults undergoing the Naturalization process informed their research project. In addition, such direct contact allows for a better understanding of the range of literacy skills and intertwined literacy services that are needed to achieve the life goal of becoming a US Citizen. The ability to view the larger picture is an important step in laying the foundation for future opportunities to further integrate academic insights with community needs, fostering ongoing mutual benefit and innovation in our endeavors.

The collaborative research initiative with Dr. Baniya and her team has brought invaluable benefits to both our learners and volunteer instructors within the citizenship preparation classes. For our learners, being actively engaged in an academic research project has instilled a profound sense of value and inclusion. By seeing their perspectives and lived experiences incorporated into the research process, they feel empowered, knowing that their voices contribute to shaping resources that can assist future applicants in navigating the complex naturalization journey. Moreover, the organic outcome of creating spaces for interaction between the research team and our learners has been transformative. Through open dialogue and attentive listening, the research team has sensitively incorporated the feedback and needs expressed by the learners into their work. In contrast to the often opaque and confusing nature of the naturalization process, where applicants may feel isolated and uninformed, this collaborative research provides a refreshing opportunity for learners to share their needs, perspectives, and concerns in a supportive environment.

From a BRL instructor's perspective, engagement in this collaborative research initiative provides a unique opportunity to contextualize their time and commitment within the larger scope of academic research aimed at facilitating the naturalization process. Many of our instructors generously volunteer their time, driven by a passion for supporting adult learners on their citizenship journey. Through their direct contact with learners and extensive teaching experience, they have gained intimate familiarity with the challenges and successes that accompany this process. Drawing upon this wealth of experiential knowledge, instructors play a pivotal role in not only delivering curriculum content but also providing crucial guidance and support to learners as they navigate the complexities of the naturalization process. Incorporating the perspectives and expertise of our instructors into the research process enriches the depth and breadth of our collective understanding. As active participants in the research process, instructors find validation in seeing their knowledge and insights recognized and utilized to drive meaningful change.

Reflection as a Methodological Grounding

In this chapter, we employ a reflective approach, drawing upon the personal experiences and perspectives of the authors involved in this digital justice project aimed at supporting refugee and immigrant communities in their pursuit of U.S. citizenship. The rationale for using reflections as the primary methodology was to capture the nuanced, contextual, and first-hand accounts of the individuals who played diverse roles in the project. This approach allowed for an in-depth exploration of the challenges, successes, and insights gained throughout the collaborative process of designing and developing a mobile application and website tailored to the needs of our target community. The benefits of this reflective approach include providing rich, qualitative data that captures the lived experiences and narratives of both ourselves as researchers and of our community partners (both PSRs and literacy center employees), offering multiple perspectives and viewpoints for a comprehensive understanding of the project's complexities, highlighting the personal growth, learning, and transformative experiences of those involved, and enabling a critical examination of the project's methods, outcomes, and implications for future community-engaged initiatives.

However, it is important to acknowledge the limitations of this methodology. Subjective biases and personal interpretations of events may influence reflections; however, being self-reflective was a way for us to ground ourselves in this project. Additionally, the findings may lack generalizability, as the reflections are specific to the context of this particular project. Despite these limitations, we believe that the reflective approach was valuable in capturing the essence of this community-engaged project, which aimed to integrate principles of social justice and participatory design in the development of digital resources for refugee and immigrant communities.

Reflections from an Undergraduate Computer Science Major (Layla)

Layla's reflections, presented in this section, offer insights into her journey as an undergraduate computer science student who served as the technical lead responsible for designing and coding the mobile application. Her narrative illuminates the transformative impact of integrating principles of social justice and community-centered design into her academic pursuits. The richness of her reflection, as well as those that follow, underscore the value of our chapter's narrative-driven approach in encapsulating the nuances and complexities inherent in community-engaged scholarship.

For our project, I served as the undergraduate technical lead, responsible for designing and coding the mobile application. My initial exposure to the intersection of technology and social justice occurred in Dr. Baniya's technical writing class, which emphasized service learning. The class focused on creating user documentation for refugees and immigrants within Virginia Tech who have low digital literacy skills. This experience introduced me to community-centered design techniques that I might have yet to encounter. More importantly, it strengthened my belief in technology as a tool for empowerment. The significance of social justice-oriented design in technology cannot be overstated, as it advocates for inclusivity and ensures accessibility for all communities. This approach resonates deeply with my academic pursuits and my commitment to utilizing my computer science skills for societal benefit.

My conviction in the transformative power of technology was further reinforced when Dr. Baniya invited me to contribute to the ACLS-funded Digital Justice Project by developing a mobile application. This involvement provided me with the opportunity to engage directly with the refugee community through our partnership with Blue Ridge Literacy. Collaborating closely with learners navigating their immigration and naturalization processes, I immediately felt a connection despite my limited personal experience with immigration. This connection was likely facilitated by my previous coursework and the fact that the learners in our focus group were all women.

This common ground proved invaluable as it nurtured a trust dynamic within the group, critical for open dialogue. It was this exchange that facilitated strong feedback, which in turn laid a solid foundation for our user-driven design approach. By actively involving the learners in the design process, we not only harnessed their insights but also honored their voices, ensuring the final product is reflective of their needs and aspirations.

Reflecting on the experience of organizing prototype workshops, I realized the importance of community input in the design process. As a computer science major, I've been trained to prioritize functional and aesthetic aspects of program design. However, this project challenged the conventional user-designer dynamic by transforming users from passive references to active design partners. This shift was crucial to embracing iterative design—a process that aims to continuously improve the software by incorporating feedback at each iteration. This approach is vital to ensuring the design's accessibility and usability, making the application not only functional but also genuinely user-friendly for those with varying levels of digital literacy. The iterative design process was significantly enriched by the direct involvement of our users, whose contributions gathered through prototype workshops allowed us to refine our application to better meet their needs and expectations. By making our design process responsive to user feedback, we could iterate more effectively, ensuring that each version of the application was a closer fit to what our users needed and wanted.

In our project, we shifted the traditional role of users from mere feedback providers to active participants in the design process, beginning with the creation of their wireframes for the application. This hands-on involvement not only democratized our design process but also ensured that the resulting application would align with the users' actual needs and preferences. The learners' wireframes provided simple yet clear outlines of desired features, making my role as a designer far more straightforward. I compiled these outlines into a cohesive list of features and interface requirements, which laid the groundwork for the subsequent stages of development.

With the user-inspired wireframes as a guide, I transitioned to coding the application. While the coding phase naturally limited user involvement, we maintained their engagement by bringing the initial version of the app back to the focus groups for feedback. The insights we received were instrumental in shaping the final design, affirming the value of continuous user participation.

These workshops, and the design process as a whole, were significantly informed by the principles of empathy, clear communication, and audience engagement that I had previously honed in Dr. Baniya's user documentation course. This foundation was crucial in creating an inclusive, informative, and collaborative environment.

This project represented a significant personal and professional milestone for me. Before this, I had never built a mobile application, despite possessing the necessary computer science skills. However, without the invaluable input and collaboration of our community partners, I doubt I would have developed the requisite design skills. This venture allowed me to undertake a considerable project, pushing the boundaries of what I believed I was capable of at the outset. Witnessing the application transition from concept to a tool in the hands of the very users who inspired its features has been profoundly rewarding. It fills me with immense pride to reflect on my growth—not only in terms of technical abilities but also through the relationships forged and the potential impact on the lives of others. This experience has underscored the true essence of technology's role in empowerment and community service.

Reflecting on the entire experience, I have come to understand that social justice-oriented design is fundamentally about building community, deeply understanding its unique challenges, and collaboratively crafting solutions. The lessons I have learned from this project will be a guiding force in my ongoing efforts to create technology that acts as a bridge to social equity, not a barrier.

Reflections from an Interdisciplinary Engineering Graduate Student (Maggie)

Offering the perspective of an interdisciplinary engineering graduate student, Maggie's narrative reflection sheds light on the challenges and opportunities that arise when integrating social justice principles into technical domains like engineering. Her account illuminates the transformative impact of this community-engaged project on her learning, which required bridging her training in engineering with participatory approaches grounded in social science methodologies. This first-person narrative highlights the importance of critically examining disciplinary boundaries and institutional cultures that may hinder justice-oriented engineering education and design.

In the context of our digital justice project, my role as the lead graduate student researcher has involved managing undergraduate student coders and researchers on our team, as well as assisting Sweta and Katy with conducting research and generating publications and presentations on our findings. This role has required me to operate as an interdisciplinary researcher, bridging my previous experiences as an engineer and a social science researcher. Integrating these two spheres of my work experience has been challenging, particularly when addressing social issues with my technical engineering background.

My initial motivation for pursuing engineering stemmed from witnessing the aftermath of Hurricane Katrina in New Orleans. Observing engineers mobilize to restore critical infrastructure inspired me to study engineering with the hope of making similar social impacts with my work. However, upon entering the technical-focused engineering field and industry, I struggled to align my engineering work with my values surrounding social justice.

As I pursue a Ph.D. in engineering education, I have found a space to explore other disciplinary perspectives on engineering and have become acutely aware of the intersections between engineering and the social issues that drive my work. It is through my immersion in this project that I have gained the transformative engineering experience necessary to understand how engineering design, community engagement, social justice theory, and critical pedagogy can intersect in my future career as an interdisciplinary engineering professor. My reflection focuses on how this project has expanded my conceptualization of social justice-oriented engineering.

One particularly novel aspect of this project was our emergent participatory action approach. Our team remained open to change throughout our process, collaborating closely with our community partners and valuing the expertise of newcomers. Our research involved not only developing applications but also conducting focus groups and prototyping sessions to gather user feedback. We approached these sessions without preconceived ideas, instead dedicating time to engage with our community and learn from them. This approach allowed us to unpack their needs and how we could address them, leading to iterative improvements in our design process, such as updating consent forms and incorporating novel translation features, study modes, and culturally relevant iconography into our application.

In this project, my eyes were opened to our community's accounts and the barriers existing ICTs present to their citizenship. I also learned about gaps in my knowledge of interdisciplinary and critical engineering projects. The engineering world with which I am familiar is ruled by Gantt charts, strict deadlines, separating social issues from a more "objective" technical focus, and a love for generalization and standardization. Emergent design approaches requiring time-consuming work are considered a 'last resort' compared to quantitative efforts that promise to apply to all with efficiency. Centering the knowledge of specific groups who may not have engineering degrees or do not represent a random population is foreign. Because of this, it took significant effort to take on this project– it required stepping out of my comfort zone and volunteering time outside of my coursework. That said, without this leap, I would likely not have had the chance to work on a design project where we chose to radically shift whose expertise is highlighted in ICT design because this sort of community engagement and social justice-oriented design is in defiance of typical engineering ideals related to efficiency, order, and universality. Where some might think we 'wasted' time, we found the gaps in existing technologies that mattered most to our community while still developing widely applicable apps. Coming out of this I realize that it is in these sorts of interdisciplinary, social-justice-oriented, and participatory action-focused projects that engineering can break out of siloed thinking running counter to digital justice.

Despite the challenges, this experience has underscored the potential of interdisciplinary, social justice-oriented, and participatory action-focused projects as well as critical pedagogy to challenge siloed thinking within engineering and contribute to digital justice. As I reflect on this project, I see barriers within higher education systems and cultures, particularly within engineering departments, that hinder the adoption of interdisciplinary pedagogies of social change. These include academic departments acting in siloes, crafting specific course requirements tailored to disciplines, employing certain methodologies aimed at specific kinds of questions, and judging academic merit on disciplinary-based aims. These dominance of disciplinary-based structures and pressures often mean that engineering classes, vocabularies, and ways of thinking take precedence over new engineers learning about participatory action methods and critical theory in ICT development. Approaches such as these—grounded in non-engineering epistemics—remain largely overlooked despite their clear potential contribution to engineering research. I now realize that through my career in engineering education, I hope to create programs that encourage students to think outside of the confines of the field's conventional design theories and approaches to community engagement. We need to go further than simply hosting one-and-done design charrettes or providing scholarships to local communities in uni-directional relationships. We need to consider ourselves as designers in a constant state of learning, stay open to the ideas of communities we serve and the time it takes to do this sort of work, as well as understand when we need to call in outside expertise. I envision projects like our critical participatory action research making ideal interdisciplinary and justice-focused engineering capstone design projects as well as laying the groundwork for future project-based learning curricula related to justice-based engineering. Through my career I hope to expand dominant engineering concepts of what constitutes effective design and community engagement to align engineering more closely with social justice goals; I believe that critical participatory action research is a key to transforming engineering design pedagogy.

Reflection from Sweta

Sweta's narrative offers insights into her journey as a faculty member whose experiences teaching citizenship classes to refugee and immigrant students inspired her to pursue this community-engaged digital justice project. Sweta's account also highlights the convergence of her background in technical communication with her collaborator Katy's extensive community partnership experience - a synergy that laid the foundation for undertaking this participatory, justice-oriented endeavor. Her reflections underscore the challenges and importance of prioritizing community voices and embedding their lived experiences into the design process.

In the fall of 2021, I volunteered for Literacy New River Valley (LNRV) in Blacksburg, Virginia, upon joining Virginia Tech as an Assistant Professor. I enrolled as a teacher for the U.S. Citizenship course designed for refugee and immigrant students, co-teaching with John Hess to a cohort of approximately nine students. Among them were four female students with whom I frequently interacted. An intriguing observation was their use of technology, particularly WhatsApp, to record my recitations and share them among themselves. This practice was instrumental in their preparation for the citizenship test, as they used these voice notes for memorization. This experience highlighted the contextual, cultural, and social dimensions of technology use.

A semester later, in a conversation with Katy, we discussed the potential of leveraging technology to address the ICT needs of the community. This discussion resonated with me, especially considering my earlier encounter with Aakash Gautam et al.'s (2020) work on developing a mobile application in collaboration with Nepali women rescued from brothels in India. The participatory approach employed in this project, involving the community in the design process, greatly intrigued me. Upon learning about the ACLS Digital Justice fund and considering the history of partnership with the community, their ICT needs, and my teaching experiences, Katy and I saw an opportunity to apply for the grant. This initiative represents a convergence of our ideas and experiences, aligning with the principles of participatory design and community engagement in technology development.

When we wrote the ACLS grant, it was a dream that we both dreamt together for the community that we wanted to serve, and the mobile application and a website were ways that we thought would be a tangible way to serve the community. As a new implant to Blacksburg, Katy provided a lot of contextual information about the work she had already been doing within the community. That work became a grounding for us to imagine collaborative work within the community. My background in technical communication combined with Katy's years of community work led us to do this collaborative project. But as Katy mentioned, this sort of collaborative work is difficult, as we were involving the community members in the design process. To incorporate refugee experiences, we conducted focus groups, prototyping workshops, user experience testing, and finally training the participants. Our first experience in focus groups led us to rethink our project in how we provide justice to women who were sharing with us their stories of hardships with the U.S. citizenship process. That led us to be very careful about how we are approaching the project and how we are working collaboratively for data collection. This is where we are indebted to Ahoo and her team at Blue Ridge Literacy. With Ahoo's support in organizing the meetings and workshops, translating the information, and most importantly, her team being there, we were better able to create a meaningful engagement opportunity. It was joyous to see how women were actively participating in the online as well as in-person meetings and workshops that we organized. We also partnered a lot with Jonathan Bradley from the University libraries who provided a lot of consultation regarding the application design, workshop, and breaking down the technological information for us. Later on, for leading our usability testing, we had another level of support from the University libraries where Kayla McNabb and her team provided support in organizing the workshop. Along with these workshops, our research assistants, Christeana Williams, Ashlyn East, Chris Jayasinghe, and Eric Kim were the students who developed the website and all the content for the website.

Therefore, this project exemplifies a collaborative effort involving university faculty, students, community organizers, and community members, all working together in the design of a mobile application. This collaboration demonstrates how justice can be integrated into the design process and highlights the feasibility of involving the community to achieve a justice-oriented design approach.

Reflection from Katy

Last, Katy's reflection highlights how Sweta's teaching experiences mirrored issues she had previously identified in her research on digital services for refugee and immigrant communities. This motivated the pair to design a participatory project that centralized immigrant voices in shaping a beneficial technological solution. In particular, Katy underscores the challenges of dedicating substantial time for meaningful community engagement, noting the critical role of institutional support structures in enabling this difficult but impactful work.

When Sweta approached me about the class she was teaching, I knew right away that her work was an example of the issues that Rebecca Hester and I had identified when working with immigrant communities. We understood from our participants that many services, including digital services, were not designed with refugees or migrants in mind. Sweta's experiences suggested that there was an opportunity to put into action a proposed project that spoke to both our research conclusions. We therefore designed our project in a way that would not only account for the immigrant experience but also would ask participants to help shape the design that would be most beneficial to them.

Over the last two years, as we've been working on this project, it has been challenging yet gratifying to see this approach come to fruition. The challenge is the time it takes to meaningfully work with participants. Making time, traveling to their spaces of comfort (i.e., Blue Ridge Literacy Center), and managing the schedule to accommodate participants, researchers, students, community literacy professionals, and design experts is complex, time-consuming, and sometimes overwhelming. However, the time this complexity takes has made for an application that is meaningful and tangibly helpful to users. I have thought often about how this work is difficult for anyone, but especially colleagues who must publish before earning tenure. This project has taken, all told, more than six years, two of which include the fieldwork that Sweta and our team have done together, not counting the writing of several grant proposals. The Center for Refugee, Migrant, and Displacement Studies has served as the infrastructure to do this work and that has made an enormous impact for me to be able to devote this kind of necessary time for the coordination, fiscal management, and writing this kind of work takes. Since our center was founded in 2020, some of the work was done before this infrastructure became a dedicated space. Many of us are doing this kind of humanities-based work outside of this kind of infrastructure. Now that I've done the work in both spaces, I appreciate even more how critical the infrastructure is. Having our institutional administrators understand and support this necessity will continue to be a critical piece as we continue to work toward bringing scholarly and experiential knowledge together to build meaningful and justice-oriented digital design for communities.

Conclusion: Incorporating Refugee Experience and Digital Literacy Conditions in Design

In conclusion, our project represents a convergence of personal, professional, and community experiences, underscoring our commitment to designing for justice. By grounding our work in an understanding of the human experience of displacement, we actively engaged with the community, recognizing their knowledge as a valuable asset in the design process of the mobile application. This approach is rooted in a humanistic framework, aiming to address the needs and integrate the knowledge of individuals seeking refuge.

Central to our participatory ICT development methodology was the involvement of refugee and migrant communities in decision-making and design processes, allowing us to incorporate their lived experiences and digital literacy capacities into the development of these digital products. Our collaboration sought to co-create digital resources that address the officer-dependent nature of citizenship processes and exams, as well as the knowledge-related and digital injustices faced by migrants in the naturalization process. Through this collaboration, we also facilitated an exchange of resources and knowledge regarding US culture, history, and politics, integrating them into tangible platforms such as the mobile application and website. Our ultimate goal is to advance equity and justice for resettled communities through these initiatives.

In this book chapter, we use narrative reflections from key project participants to emphasize the importance of iterative design based on community engagement as well as on the challenges of traditional top-down design methodologies in ICT development. Overall, the reflections add to our understanding of social justice work in ICT development in several ways. They emphasize the importance of acknowledging and overcoming siloed thinking within disciplines like engineering, highlight the value of fostering collaboration between academics and community partners throughout the design process, and showcase the transformative potential of participatory action research in ensuring that ICT solutions are truly responsive to the needs and experiences of marginalized communities.

While our applications serve specific purposes, we hope that sharing our experiences can inspire faculty, students, and community members to consider how they can contribute to a broader understanding of the digital divide and its impact on refugee resettlement in the US. By sharing our journey of learning, sharing, and collaborative work, we aim to encourage others to engage in similar efforts to promote justice and equity in their respective communities.

References

Alam, K., & Imran, S. (2015). The digital divide and social inclusion among refugee migrants: A case in regional Australia. *Information Technology & People*, 28(2), 344–365.

Allen, B. L. (2018). Strongly participatory science and knowledge justice in an environmentally contested region. *Science, Technology, and Human Values*, 43(6), 947–971. https://doi.org/10.1177/0162243918758380

Baniya, S. (2022). Rethinking access: Recognizing privileges and positionalities in building community literacy. *Community Literacy Journal*, 17(1), 50–65.

Baniya, S., Brein, A., & Call, K. (2021). International service learning in technical communication during a global pandemic. *Programmatic Perspective*, 12(2), 26–58.

Baniya, S., Brien, A., Call, K., & Kumar, R. (2022). Covid, international partnerships, and the possibility of equity: Enhancing digital literacy in rural Nepal amid a pandemic. *Reflections: A Journal of Community-Engaged Writing and Rhetoric*, 21(1). https://reflectionsjournal.net/2022/02/covid-19-international-partnerships-and-the-possibility-of-equity-enhancing-digital-literacy-in-rural-nepal-amid-a-pandemic/

Bletscher, C. (2020). Communication technology and social integration: Access and use of communication technologies among Floridian resettled refugees. *Journal of International Migration and Integration*, 21, 431–451.

Fedyuk, O., & Zentai, V. (2018). The interview in migration studies: A step towards a dialogue and knowledge co-production? In R. Zapata-Barrero, & E. Yalaz (Eds.), *Qualitative research in European migration studies* (pp. 171–188). Springer International.

Gautam, A., Harrison, S., & Tatar, D. (2020). Crafting, communality, and computing: Building on existing strengths to support a vulnerable population. In *Proceedings of the 2020 CHI Conference on Human Factors in Computing Systems*. http://aakash.xyz/files/chi2020.pdf

Hester, R., Powell, K. M., & Randall, K. (2021). Refugee and migrant partnerships in Virginia: Fostering connectivity and reciprocity. In M. Cowell & S. Lyon-Hill (Eds.), *Vibrant Virginia: Engaging the Commonwealth to expand economic vitality*. Virginia Tech Publishing.

Jones, N. N. (2016). Narrative inquiry in human-centered design: Examining silence and voice to promote social justice in design scenarios. *Journal of Technical Writing and Communication, 46*(4), 471–492. https://doi.org/10.1177/0047281616653489

McCaffrey, K. T., & Taha, M. C. (2019). Rethinking the digital divide: Smartphones as translanguaging tools among Middle Eastern refugees in New Jersey. *Annals of Anthropological Practice, 43*(2), 26–38. https://doi.org/10.1111/napa.12126

Pourchot, G., Hassouna, K., Powell, K. M., Randall, K., Shadle, B., & Swarup, S. (2018). *Refugee integration: Data for smart policy (policy brief)*. School of Public and International Affairs, VT.

Powell, K. M., & Hester, R. J. (2018). *Understanding information and communication technologies through resettlement narratives: Defining "epistemic integration" for knowledge justice*. White Paper. Virginia Tech Center for Refugee, Migrant, and Displacement Studies.

Powell, K. M., Hester, R. J., & Randall, K. (2019). Refugee and migrant partnerships in Virginia: Fostering connectivity and reciprocity (Report from the Fostering Reciprocity Symposium in Arlington, VA). Virginia Tech.

PART II

Community-Engaged
Design Efforts in Action

7

BEYOND "MAINTAINING STATUS"

A Call for Distributed Responsibility in the Professionalization of International Graduate Scholars

Felicita Arzu-Carmichael, Therese I. Pennell, and Josephine Walwema

Beyond "Maintaining Status": A Call for Distributed Responsibility in the Professionalization of International Graduate Scholars

Design thinking is both a method and a way of thinking about innovation with users in mind. It uses an iterative process of empathize, define, ideate, prototype, and test (Kimbell, 2011; Lockwood, 2010) and is a "...delicate process to assuring justice and avoidance of harm" (Lane & Moore, 2023, p. 31). In technical and professional communication (TPC), design thinking has become a rapidly growing area of scholarship, valued for its approach to innovation and human-centeredness (Tham, 2019; Verhulsdonck et al., 2021). Its third phase, ideating, is critical because it offers possible solutions before settling on an ideal solution. In the context of this article, we focus on finding ideal professionalization solutions that are attentive to the material needs of international graduate scholars (IGS) and that are rooted in the community. Community is defined here based on the ideas of communities of practice in which members have a shared interest in the upliftment/betterment of the group. Universities can be considered a cluster of communities, with each program, college, and administrative office functioning as a community. For IGS, families and friends are often in faraway countries so academia is not only where we are educated and work but also where our peers and faculty become friends and family. In this light, we acknowledge the community with whom we engage in design work is the university. As faculty members, we are aware of the siloing in the university community, which creates obstacles for professionalizing IGS. Too

DOI: 10.4324/9781003469995-10

often understanding and sharing knowledge of immigration policy, for example, the international student services office does not communicate with graduate programs who do not communicate with English programs, and so on. And yet each of these communities significantly influences the success of IGS.

The rubric of iterative design offers a process through which we as scholars and designated school officials (DSO), namely faculty, program administrators, and the office of international students (ISO) as a community can devise processes to better respond to the immigration and rhetorical needs for international scholars. These revisions would require the collective participation of a variety of campus entities because matters concerning immigration affect everyone in some way, not only international scholars.

The existence of the Presidents' Alliance on Higher Education and Immigration, a national organization founded in 2017 and composed of university presidents and chancellors, seems to suggest that immigration is, in fact, a university-wide issue. Its members address issues regarding higher education and immigration policies and practices that impact university students, including undocumented or Deferred Action for Childhood Arrivals (DACA, a program that protects persons from deportation) students, international students, refugee and other immigrant students, and immigrant and international alumni, faculty, and staff (Presidents' Alliance on Higher Education and Immigration, 2023). Yet, the professionalization of IGS often seems to be separate from the work that occurs in the office of international students.

For most graduate professionalization programs, the focus is on fundamental components necessary to secure an academic position, such as preparing a CV and cover letter; applying for and interviewing for academic positions; as well as expectations pertaining to types of universities, and academic positions, among others. Indeed, most programs dedicate time ranging from a day-long workshop to a series of workshops to professionalization. In focusing on these fundamentals of the academic job market, professionalization programs fail to *mind the gap* between IGS, who are constrained and circumscribed by various immigration and visa protocols, and their non-immigrant counterparts. Treating all students equally means that only the scholars affected by immigration are paying attention to the various deadlines and tasks necessary for them to successfully complete their program(s) of study and maintain status. Thinking linearly about this process does not account for contingencies and material needs related to immigration laws, specific program completion deadlines, and job market orientation.

Therefore, this chapter aims to disrupt that status quo by addressing the following question: How can iterative design bring about change in graduate professionalization programs through "distributed responsibility"? To answer this question, we bring together scholarship on design thinking, compliance, and Black feminist epistemologies to address the IGS gap in professionalization and to collectively describe a theory of design shaped by the needs and experiences of international scholars. We draw parallels among our collective experiences as international scholars and Black diasporic women who have undergone this process.

In the sections that follow, we first interrogate professionalization as currently designed and practiced in academia. We then demonstrate how professionalization practices in academia are shrouded in an ethic of compliance that not only leaves a gaping chasm for IGS but imbued with questionable ethics. Next, drawing on Black feminist epistemologies, we propose that design thinking rooted in iterative design allows for what we are calling *distributed responsibility*, a mechanism for programs to collectively engage in more inclusive and ethical approaches to professionalization that mind the needs of all graduate students, their immigration status notwithstanding. Community engagement on university campuses, including, but not limited to, international student services offices, employee and student resource groups, the office of legal affairs, graduate programs, the office of graduate studies, and what the Department of Homeland Security calls, "designated school officials," are all implicated in this community of distributed responsibility. We conclude by calling for distributed responsibility as a component of the professionalization of international graduate scholars.

Professionalization and Its Discontents in Academia

The process of professionalizing individuals is a contested site for some disciplines. And this is a good thing. Programs have been tasked to not only train but to develop a body of knowledge that makes its members' abilities more exclusive from other disciplines. In technical communication, for example, researchers have consistently scrutinized technical communication programs and questioned whether they meet the needs of communicators in the field and whether they can improve market conditions (see, for example, Carliner, 1996, 2012; Savage, 1999, 2013). Such scrutiny of programs hones the professionalization of individuals allowing for robust content that is dynamic. Hart (2008) describes professionalization as having "an incredible predictability and normalizing power" (p. 203) so much so that university administrations and faculty members view professionalization as key to establishing equitable environments for diverse individuals. Hall (1968) explains that professionalization "strengthens attitudes" of its members (p. 204).

Professionalization can encompass a wide range of ideas. For clarity, we focus on two aspects: training of graduate students and professionalization of new faculty. For graduate students, programs provide training for the academic job market, these include activities like completing mock job interviews, understanding the publication process, finding and tracking open positions, putting together the job application package, and completing the campus visit. The second aspect of the professionalization practices we interrogate is the professionalization practices for newly hired faculty. These practices include helping new faculty understand their specific universities' tenure-track process: how teaching, service, and research are weighted/valued to successfully secure one's position.

While there is no standard procedure, a cursory review of several graduate programs reflects a perceptible approach to the process. UC Berkeley Graduate Division, for example, has developed a set of "core competencies" for its graduate student's professional development guide that lists a series of "development workshops and events" six in total through which graduate students can be professionalized. Berkeley Career Engagement also has a page devoted to "the transition from Grad Student to assistant professor" which outlines the nature of academic positions and what the duties and responsibilities entail. At Harvard, there is a page devoted to "The PhD Problem" (Menad, 2009), while on the MLA association website's page on professionalization, the association lays out a statement "in perspective" and then outlines five tenets of professionalization (which it says is the purview of PhD granting institutions) to focus on enculturating students into the profession, informational, mentoring, pedagogy, and asking students to take charge "in their own professional preparation" (para. 13). The MLA sees its role as mostly advisory in "urging" institutions to take up professionalism. What these professionalization workshops/programs reflect is a "linear and orderly progression" to professionalization that fails to mind the gap for IGS.

Carliner (2012) categorized how professionalization works into three categories: formal, quasi, and contra-professionalization. Formal professionalization refers to practices that fully support individuals, quasi professionalization refers to practices where individuals have to fill in gaps left by their programs, but that they need to function in their occupation; and contra-professionalization, the most problematic because, as Carliner (2012) describes it, it refers to initiatives where individuals go "outside of parts of or the entire framework of the profession or occupation" (p. 55). That is, contra is the least supportive form of professionalization. How well do our universities assist graduate students and new faculty in navigating the US immigration policy in its professionalizing activities? An element of contra-professionalization for IGS is the lack of discussion of immigration policies that dictate the fate of too many members of, if not all, university programs.

It is no secret that the benefits of professionalization depend on identity. For women, minorities, people with disabilities, and immigrant individuals, professionalization and academic hiring activities either overlook or come up lacking in helping them secure a position and/or feel secure in the profession. Pennell et al. (2018) found that "most of the participants [in their study] experienced difficulties transitioning from their PhD programs to life as faculty members" and that participants' identity characteristics forced them to go outside of established formal professionalization activities to get needed training. Kumar (2020) describes higher education's attempt to recruit international graduate students as diversity tactics and its ineptness in training these students for the job market as "diversity that does not cater to the needs of the diverse people [which] is performative at best and exploitative at worst" (para. 3). As well, Walwema and Arzu Carmichael's (2021) investigation of the hiring practices, more specifically, the job descriptions in academia (specifically, rhetoric and composition and technical communication) found they can be exclusionary to immigrant applicants. For Chen et al. (2022), "centering marginalized perspectives may be conducive for imagining hiring practices that are humane, equitable, and accessible" (p. 21).

While Kumar (2020) celebrates her graduate education, her criticism is laid bare at the feet of how institutions seem to sacrifice their recruited international students and immigrant faculty at the altar of US immigration policies. Based on our reading of the research, programs insufficiently scrutinize or are oblivious to immigration policy, the same programs that house international students and faculty. Outside of rhetoric, composition, and technical communication, the sciences face significant recruitment and retention issues because of current visa and immigration policies. Feeney et al. (2023) explain that the US immigration and visa policy "negatively impact[s] diversity, attraction of talent, open science, competitiveness, scientific workforce development, and strength of US technological industries" (p. 3). With the challenge of American universities in recruiting, hiring, and retaining IGS, one must ask, how do these institutions grapple with the problem? Based on the findings of Walwema and Arzu Carmichael (2021) as well as the goal of Chen et al. (2022), academic institutions seem to have a reactive attitude that does not take into consideration the humanity of their students and employees.

Understanding the US immigration policy may seem like an unimportant practice that focuses on a small percentage of individuals in any program. However, IGS are an established tradition in academia. The creation of The Presidents' Alliance on Higher Education and Immigration underscores the importance of IGS in academia. As noted by Feeney et al. (2023) and Kumar (2020) "higher education" has been negatively impacted by US policies and, with this, the exchange of knowledge globally is hindered by restrictive immigration policies.

For IGS, support services often focus on maintaining one's immigration status. This means staying within the guidelines of the F-1 student visa for international students, securing and staying within the guidelines of Optional Practical Training for temporary employment as a student, and transitioning to the H1B visa, a non-immigrant visa for persons who work in a specialty occupation. Each of these statuses has predetermined activities that must be completed by IGS, for example, international students on an F1 visa are not allowed to work outside of university employment, and hours are also limited to a maximum of 20 hours. IGS also must be aware of the various US agencies that they need to comply with, the United States Citizenship and Immigration Services (USCIS), the Department of Homeland Security (DHS), and customs and border patrol (CBP) all influence some aspects of the lives of IGS. In addition, each immigration status dictates a limited time for holders to stay within the country, one must be aware of their program's end date (PED). If a graduate student's study is delayed for any reason, they must file an extension that must be approved by their program, the university, then the federal department in charge. Timeliness is key. IGS must keep track of these dates, the limitations of their status, and the agency that they must contact. Missing a date, and completing any activity that falls outside of their status can mean losing their status, being deported, and often being restricted from ever returning to the US.

As institutions, universities have more resources than individual students and faculty to interpret immigration laws that are constantly shifting (for example, between 2017 and 2021 the Trump Administration instituted 472 immigration policies (Bolter et al., 2022) and policies change as individuals change status, for example, from student to faculty). It is not only important that universities help IGS "maintain their status"; it is imperative that a community of DSO's and faculty be kept in the loop as well.

All three of the authors of this chapter accept that professionalization practices within programs are incredibly important. We also concede that they are lacking in some respects. For IGS to have to hunt for pertinent information outside their respective institutions so as to continue in their occupations reveals a significant gap and speaks to the contra-professionalization Carliner (2012) writes about. As Harper and Kezar (2023) remind us, programs tend to be designed based on the expertise of staff and what they believe to be the needs of students. We believe this is true of academic programs, too. IGS are often locked out from opportunities that fall outside of immigration stipulations and are required to find out about, understand, keep up with, and financially afford US immigration policy requirements which is near impossible at times. It is important, then, that universities reconsider the design of their professionalization training to eschew compliance rhetoric.

Compliance, Professionalization, and IGS

We have outlined the generic nature of professionalization and how it is mainly concerned with compliance. As long as programs can offer students the steps in the generic academic job application, they might consider themselves in compliance with professional expectations.

Compliance rhetoric is associated with control of information or of one by another and can occur anywhere from interpersonal relationships to cognitive and affective ways of "gaining strategies" (Martins, 2005, p. 62). In its earliest iteration, compliance rhetoric gained a lot of traction in medical situations, which (Segal, 1993) ascribed to the inadequacy of persuasion in science (p. 91). The mechanism of patient compliance, Skinner and Franz (2018) found, conveyed "medical authority and power" (p. 240) and with it, paternalism and lack of patient autonomy, undergirded by the lack of health literacy among patients.

For our purposes, compliance is always associated with conforming unquestioningly to systems, processes, and procedures—to maintain status. Understanding compliance through a Foucaultian lens, Skinner and Franz (2018) see in compliance the twin concepts of power/knowledge in which forms of knowledge are present in binaries such as "as normal/abnormal" or, in our case, in/out of status. These binaries then make a typical approach to compliance inevitable, so that knowledge about systems and processes, for example, becomes commodified to be wielded with authority, and inevitably control. Compliance rhetorics leads naturally to a sort of pliable conformity to systems of power by those (dis)affected.

However, beyond compliance is a paradigm that centers communities and their ways of knowing. In this paradigm communities "serve as experts … due to their intimate knowledge of local norms and practices" (Skinner & Franz, 2018, p. 253). Community knowledge emerges from and through discursive practices such as that in which the "rhetorics of self" surface within similarly affected individuals—as is the case for the authors of this chapter. It requires that those in power allow for the voices of affected individuals to speak to and define the conditions as seen through their lived experiences. As these epistemologies of self are given room to surface, they shift the ideation of solutions so that a more equitable and just system begins to replace the formerly skewed model of dominance and compliance and, ultimately, punishment.

In immigration rhetorics, the need for compliance with regulations is at the forefront of maintaining immigration status. But beyond maintaining immigration status, affected individuals, in this case, IGS, have learned to thrive based on knowledge gained from their lived experiences on visa status. Such knowledge can be leveraged in a distributed community of university DSOs responsible for their success. Prioritizing IGS needs, these DSOs can incorporate that knowledge and deploy it throughout the professionalization practices in graduate programs. It is a slow iterative-design-led approach fashioned within the ethics of care, specifically, Black Feminist Care that would result in inclusive and equitable practice.

Rather than operate within a compliance mindset, we agree with Costanza-Chock (2018) that the "procedural and distributive" ethic of design thinking to "advance the participation of marginalized communities in … the design process" is an ethical imperative (p. 534). Imagine how IGS can best be served by bringing together a community of campus DSOs, each bringing a diverse set of expertise, positionality, and power to fundamentally alter the professionalization of IGS. Such a justice-oriented design of professionalization would account for contingencies in the lives of IGS, students with disabilities, and those beyond the binary.

Black Feminist Epistemology and Iterative Design

This project was inspired by our experiences as Black immigrant women academics; thus, we draw on Black feminist thinking because as Barbara Smith (1985) affirms, "…Black feminism is, on every level, organic to Black experience" (p. 4). Black feminism recognizes that the major systems of oppression are interlocking (Collins, 2009), and it "prioritizes action, experiences, and epistemological frameworks beyond the theoretical" (Moore, 2018, p. 188). Therefore, we are compelled to draw on a Black feminist epistemology to help us identify the ways in which systems of oppression intersect and affect immigrant scholars, who exist at the intersections of race, class, gender, and national origin. Black feminist epistemology also recognizes that "U.S. Black feminism participates in a larger context of social justice that transcends U.S. borders" (Collins, 2009, p. xiii) particularly as it relates to women of African ancestry. As TPC continues to advocate for social justice and for the voices and perspectives of marginalized populations to be centered (Savage & Dura, 2023; Walton et al., 2019), this chapter calls attention to the immigration concerns of IGS to contribute to these advocacy efforts.

In "Some Home Truths on the Contemporary Black Feminist Movement," Barbara Smith (1985) argues that "A commitment to principled coalitions, based not upon expediency, but upon our actual need for each other is a second major contribution of Black feminist struggle" (p. 9). Ideas of such principled coalitions in Black feminist thinking intersect with recent calls for social justice advocacy and coalition work within TPC. For example, Walton et al.'s (2019) 4R heuristics (recognize, reveal, reject, and replace) promote coalitional action that aims to bring about equity and inclusion for multiple marginalized populations. In terms of the Black feminist struggle, Smith (1985) indicates that coalitional thinking and Black feminist theory specifically "clarifies the nature of Black women's experience, makes possible positive support from other Black women, and encourages political action that will change the very system that has put us down" (Smith, 1985, p. 10). Importantly, coalitional action is different from support. In distinguishing between solidarity and support, Bell Hooks (1984) has written that "To experience solidarity, we must have a community of interests, shared beliefs and goals around which to unite, to build Sisterhood. Support can be occasional. It can be given and just as easily withdrawn. Solidarity requires, on the other hand, sustained, ongoing commitment (Hooks, 1984, p. 64). This solidarity and coalitional work that is necessary for the equity and inclusion of Black women and other marginalized groups is a fundamental aspect of our understanding of design thinking and knowledge-making.

Black feminist theory also requires making the distinction between knowledge and wisdom, which Hill Collins (2009) notes is a distinction that "has been key to Black women's survival" (p. 276). Thus, we draw on Hill Collins's tenets of Black Feminist Epistemology to help us develop a critical understanding of advocacy for international scholars, the multiplicity of oppression, and programmatic design. The first dimension of Hill Collins's Black feminist epistemology is *lived experience as a criterion of meaning*. The issues that we study are not separate from the experiences we have. In other words, our knowledge is connected to our lived experiences. Hill Collins takes this idea a step further arguing that lived experience is the factor that allows Black women to be able to distinguish between knowledge and wisdom, a distinction that is fundamental to our survival. Next, *the use of dialogue in assessing knowledge claims* is another tenet of Hill Collins's epistemology. Hill Collins indicates that "For Black women new knowledge claims are rarely worked out in isolation from other individuals and are usually developed through dialogues with other members of a community" (p. 279). In fact, this community work in developing new knowledge is deeply rooted in "African-based oral traditions and in African-American culture" (p. 279). Furthermore, knowledge is intertwined with *the ethics of caring*, which is Hill Collins's third characteristic of Black feminist epistemology. Our "personal, expressiveness, emotions, and empathy are central to the knowledge validation process" (pp. 281–282). Hill Collins' additional key components of the ethics of caring, which include individual expressiveness (individual uniqueness enriches the group), appropriateness of emotions (emotions are connected to intellect), and the capacity for empathy is key. Like all other tenets of Black feminist thinking, these ideas challenge traditional claims about how knowledge is formed. Essentially, Black feminist epistemology allows us to "advance new, alternative, and sometimes oppositional interpretations" (Henry, 2005, p. 96) that might shift how we think about our experiences. Finally, Black feminist thinking requires that we are accountable for our knowledge claims. Thus, *the ethic of personal accountability* means evaluating an individual's character, values, and ethics (p. 284) by identifying any evidence connected to their credibility as an ethical human being.

These four tenets of Black feminist thinking provide us with a knowledge base for iterative design. As theorized by Lane and Moore (2023), iterative design functions as an intersectional feminist methodology. Lane and Moore, who draw from Hill Collins's (1990) work, affirm that there are three inherent values of iterative processes: non-linearity, slowness, and multivocal critical imagination. They also affirm that "engaging in iterative design as non-linear, slow, feminist work can help ward off injustices" (p. 41). Lane and Moore (2023) remind us that "We mirror the ethos of the organizations with which we are involved, which often values swift production and delivery of various goods—products, research, ideas. Yet with the growing popularity of design thinking and ideation, a recent movement toward slow, deliberate moves of production have emerged, more directly valuing a socially just and resistant orientation toward wicked problems" (p. 39).

As we noted in the introduction to this chapter, most graduate professionalization programs tend to focus on components necessary to secure an academic position, such as preparing a CV and cover letter; applying for and interviewing for academic positions; as well as expectations pertaining to being on the academic job market. If such programs were to engage in the non-linear, slow, and multivocal work in designing their programs, we argue, the experiences of IGSs would not be fraught with quasi and contra-professionalization forays as they try to successfully complete their program(s) of study while also maintaining status.

Moreover, Lane and Moore's views on *iterative design as non-linear* and *slow work* and Hill Collins's use of dialogue and ethics of caring tenets help expand our approach to design thinking. For example, TPC's critique of expediency (Katz, 1992; Lane & Moore, 2023) highlights the importance of slow work because it brings to light the inherent harmful consequences within. This "slow valuation" (Lane & Moore, 2023, p. 38), we believe, is not removed from the "personal, emotions, and empathy [that] are central to the knowledge validation processes" (pp. 281–282). And Cooper (2018), on the importance of emotion, argues that "We (feminists) need to embrace our messiness more. We need to embrace the ways we are in process more" (pp. 5–6).

Finally, from a Black feminist perspective, *multivocality* recognizes the importance of *dialogue in assessing knowledge* claims and adhering to the perspectives of multiple voices; it also recognizes the individual uniqueness that one brings to the group situation. Importantly, Lane and Moore (2023) argue that iterative design "often (though not always) insists upon knowledge-making practices that eschew the certainty of a singular source of expertise and truth" (p. 39). Our own dialogue as we collaborated on this chapter revealed a connectedness that was only possible through our frequent dialogue. Therefore, collective practices are valued while being mindful of the unique positionality of each contributor.

Applying Black Feminist Epistemology and Iterative Design to Professionalization: A Framework for Distributed Responsibility

When international students are accepted into graduate programs, their mere presence does not guarantee equitable and inclusive professionalization practices. As we noted in the introduction, graduate programs tend to privilege particular groups in their design and policies, middle-class able-bodied, white cisgender men, who are native English speakers and have U.S. citizenship, an approach which further harms multiply-marginalized groups such as immigrants. Yet, this conventionality has largely gone unquestioned. One need only to read Kimberlé Crenshaw (1989)'s "Demarginalizing the intersection of race and sex: A black feminist critique of antidiscrimination doctrine, feminist theory and antiracist politics" to understand the harmful consequences that impact Black women specifically, and marginalized populations more broadly, when the intersectional experiences of these groups are not factored into policies and practices.

Moreover, Kynard (2015) argued that "The institutional racism in which students and faculty must daily think and act is always very real and moving according to the specificity of two directions: the local situation and the national tenor of the moment" (p. 2). During times of problematic public discourse surrounding immigrants and immigration policies, the national tenor surrounding immigration becomes alarming. International scholars are accepted into graduate programs but their professional training and job-seeking prospects are fraught with obstacles. In a previous study Walwema and Arzu Carmichael (2021) identified how xenophobia manifests in job application portals and the academic job market process.

In identifying connections between design thinking and exclusionary practices that impact international scholars, we wish to develop an activist approach to graduate program professionalization that accounts for intersectional oppression that many multiply marginalized international scholars face. We aim to "find more and better ways to listen to the multidimensional voices that are speaking from within and across many of the lines that might divide us as language users—by social and political hierarchies, geography, material circumstances, ideologies, time and space, and the like" (Royster & Kirsch, 2012, p. 4). Thus, in the course of writing this chapter, we discussed ways in which we can draw on our collective experiences to investigate these matters in hopes of enacting change. In what follows, we articulate a design of what we call "distributed responsibility." In this distributed responsibility framework, we draw on iterative design to offer possibilities of (re)shaping graduate programs.

Our framework of a distributed responsibility also necessitates a rethinking of what career preparation looks like for IGS. For example, the University of Michigan offers an online course titled "Preparing for Graduate Study in the U.S.: A course for international students" to support graduate students in pursuing their degrees. Such a course can be reimagined to address the specific professionalization needs of IGS in TPC programs. The course could also be designed with IGSs to ensure "fair and meaningful participation in design decisions" by those most impacted by it (Costanza-Chock, 2018, p. 529).

The generic professionalization processes and procedures currently in place across many institutions and designated school officials across campuses responsible for the needs of IGS should take part in this distributed responsibility framework. These entities can start by paying particular attention to the timelines that graduate students on F-1 Visas need to abide by as they simultaneously move toward completion of their studies. Variations of this framework of distributed responsibility model that we envision exist to a degree at academic institutions. They include the Disability Services Office, offices of Academic Advising, as well as Student Athletic Academic Services (SAAS) among others. Variously, these services, as scholars have noted, exist to ensure "university compliance with legal requirements" first and foremost, which goal competes with that of student support (see, for example, Tamjeed et al., 2021, p. 14). We acknowledge these shortcomings even as we suggest that a similarly structured framework of distributed responsibility be instituted to cater to the needs of IGS. The SAAS, for example, has a system that monitors the performance of student-athletes for their classes in which they are registered, which could help inform the one we envision.

Consider that the final year of graduate studies is often the dissertation phase during which graduate students conduct research, collect and analyze data, write and submit chapter drafts, and work closely with their advisor. Graduate students have to be simultaneously self-directed (by demonstrating the ability to conduct independent research) and directed by their advisors and committee members to successfully complete their dissertations. This phase further necessitates that the student prepares materials to join the academic job market. In the case of IGS who are eligible for employment, but not yet authorized to engage in such employment (see Walwema & Arzu Carmichael, 2021), additional procedures tied to predetermined timelines and program end dates on their Form I-20 have to be met. The Form I-20 is a legal document that international students studying in the United States must have for immigration and academic purposes. The office of international students at every university is responsible for interpreting DHS law as it applies to IGS. For students transitioning from the student F-1 Visa, the first step is to apply for Employment Authorization. Among the documentation required for that application is PED. This form reflects the expected date of completion also listed on the student's Form I-20. The I-20 indicates an end date by which the affected student will complete the program. Because this form has to be signed by the student advisor, often the dissertation chair, the burden of explaining its necessity lies with the student, who simultaneously has to obtain the signature while signaling to the advisor the procedural nature of the PED. The advisor has to be persuaded that signing the form does not necessarily bind the committee to signing off on a yet-to-be-completed dissertation.

This is an example of the precarity that already marginalized IGS have to engage with persons in positions of power and having to educate them about the inner workings of the DHS. For IGS, the power differential alone places the scholar in a precarious situation. A distributed responsibility framework across the community of DSOs with a vested interest in the success and professionalization of IGS that functions much like the SAAS might come in handy. For example, SAAS, which appears to be designed with a feminist epistemology: acknowledges the student's lived experience, use of dialogue in assessing knowledge, and built on ethics of care. SAAS keeps track of the progress of student-athletes through various methods, including progress reports completed by faculty, self-reporting, and academic support locations where student-athletes meet with tutors and mentors. The international student's office and the school of graduate studies already have lists of IGS. What the office of graduate studies may not have is immigration related information. To activate distributed responsibility,

- The Office of International Studies shares PED listed on the student's Form I-20 with the office of graduate studies.

- The office of graduate studies notifies students and advisors about the PED. The international graduate student signals readiness (or not) to both offices. In case the student is not on track to meet that PED, they file for an extension with the International Student Office.
- For the student on track, the office of graduate studies sends out the PED form requesting the advisor's signature.
- The advisor signs and sends back the form to the originating office, which in turn shares it with the international student office.
- The form is added to the IGS student file in the International Student Office.

This is how a distributed responsibility framework emblematic of Black feminist ethical care and that offsets feelings of isolation for IGS can be operational. It is reflective of a community of interest and is ongoing because it is stitched within professionalization protocols. It promotes a dialogic new alternative that speaks to the lived experiences of IGS without being limited to compliance and maintaining status.

Kumar (2020) in her discussion of career preparation issues that international graduate students face, singles out infrastructure challenges to adequately accommodate students' career goals. She argues that most higher education institutions are resistant to hire IGS into tenure track positions due to "visa-related complications" (para. 2). Such a finding aligns with our own analysis of job descriptions in the academic market where most institutions do not reveal whether or not they sponsor work-visas or a few indicate explicitly that they do not sponsor work visas (Walwema & Arzu Carmichael, 2021). Furthermore, each year, there are more new PhDs in English than there are new job postings (Skinnell, 2023). Given these realities, it is imperative that graduate programs rethink their practices that force IGS to engage in contra-professionalization to prepare for careers that are not academic positions (Kumar, 2020). Since IGS has work restrictions that are limited to on-campus employment and for only 20 hours a week, a more inclusive approach to professionalization would require that universities and graduate programs become educated on finding ways to have IGS secure internships off campus. And if these opportunities are unpaid, they need to determine creative ways of securing internal funding.

Conclusion

In this chapter, we articulated a framework for distributed responsibility that is design-driven and that advocates for the needs, interests, and expectations of IGS through campus community engagement couched in Black Feminist Care. We contribute to the literature on what it means when responsibility is distributed across entities in a campus community that strives for the "success" of all its students and to call attention to the possibilities of injustices in program design. To do this, we first interrogated professionalization practices in academia and demonstrated how such practices are shrouded in an ethic of compliance that does not account for IGS. We then drew on Black feminist epistemologies to inform our approach to iterative design as a theory and practice. We also relied on our own lived experiences as Black immigrant women academics to provide new possibilities for activist approaches to professionalization and to remind readers of the material impacts of professionalization practices on the bodies of immigrant scholars.

References

Bolter, J., Israel, E., & Pierce, S. (February, 2022). Four years of profound change: Immigration policy during the Trump presidency. Migration policy institute. https://www.migrationpolicy.org/research/four-years-change-immigration-trump

Carliner, S. (1996). Evolution-revolution: Toward a strategic perception of technical communication. *Technical Communication, 43*(3), 266–276.

Carliner, S. (2012). The three approaches to professionalization in technical communication. *Technical Communication, 59*(1), 49–65.

Chen, C., Bose, D. K., Sano-Franchini, J., Kirycki, E. K., Osorio, R. D., & Tetreault, E. (2022). Interrogating the four Ps: Positionality, privilege, power, and professionalism in the rhetoric and composition job market. *Composition Studies, 50*(3), 20–39.

Collins, P. H. (2009). *Another kind of public education: Race, schools, the media, and democratic possibilities.* Beacon Press.

Cooper, B. (2018). *Eloquent rage: A Black feminist discovers her superpower.* St. Martin's Press.

Costanza-Chock, S. (2018). Design justice: Towards an intersectional feminist framework for design theory and practice. *Proceedings of the Design Research Society.*

Crenshaw, K. (1989). Demarginalizing the intersection of race and sex: A black feminist critique of antidiscrimination doctrine, feminist theory and antiracist politics. *University of Chicago Legal Forum, 1989*(8). http://chicagounbound.uchicago.edu/uclf/vol1989/iss1/8

Feeney, M. K., Jung, H., Johnson, T. P., & Welch, E. W. (2023). US visa and immigration policy challenges: Explanations for faculty perceptions and intent to leave. *Research in Higher Education, 64*(7), 1031–1057. https://doi.org/10.1007/s11162-023-09731-0

Hall, R. H. (1968). Professionalization and bureaucratization. *American Sociological Review, 33*(1), 92–104. https://doi.org/10.2307/2092242

Harper, J., & Kezar, A. (2023). Designing with, not for students: Prioritizing student voice using liberatory design thinking. *About Campus, 27*(6), 31–39.

Hart, J. (2008). Mobilization among women academics: The interplay between feminism and professionalization. *NWSA Journal, 20*(1), 184–208. http://www.jstor.org/stable/40071258.

Henry, A. (2005). Chapter four: Black feminist pedagogy: Critiques and contributions. *Counterpoints, 237,* 89–105.

Hill Collins, P. (1990). *Black feminist thought: Knowledge, consciousness, and the politics of empowerment.* Unwin Hyman.

hooks, b (1984). *Feminist theory: From margin to center.* Southend Press.

Katz, S. B. (1992). The ethic of expediency: Classical rhetoric, technology, and the Holocaust. *College English, 54*(3), 255–275.

Kimbell, L. (2011). Rethinking design thinking: Part I. *Design and Culture, 3*(3), 285–306.

Kumar, K. S. (2020, April). International student precarity in the humanities academy. *Inside Higher Ed.* International Student Precarity in the Humanities Academy (insidehighered.com)

Kynard, C. (2015). Teaching while Black: Witnessing and countering disciplinary whiteness, racial violence, and university race-management. *Literacy in Composition Studies, 3*(2), 1–20. https://licsjournal.org/index.php/LiCS/article/view/939/795

Lane, L., & Moore, K. (2023). The invisible work of iterative design in addressing design injustices. *Technical Communication and Social Justice, 1*(2), 28–48.

Lockwood, T. (2010). *Design thinking: Integrating innovation, customer experience, and brand value.* Allworth Press.

Martins, D. S. (2005). Compliance rhetoric and the impoverishment of context. *Communication Theory, 15*(1), 59–77.

Menad, L. (2009, November–December). The Ph.D. problem: On the professionalization of faculty life, doctoral training, and the academy's self-renewal. *Harvard Magazine.* https://www.harvardmagazine.com/2009/11/professionalization-in-academy

Moore, K. R. (2018). Black feminist epistemology as a framework for community-based teaching: Key theoretical frameworks. In Haas, A. M., & Eble, M. F. (Eds.), *Key theoretical frameworks: Teaching technical communication in the twenty-first century* (pp. 185–211). University Press of Colorado. http://www.jstor.org/stable/j.ctv7tq4mx

Pennell, T. I., Frost, E. A., & Getto, G. (2018). Valuing contra-professionalization: Analyzing successful professionalization practices in technical and professional communication. *Programmatic Perspectives, 10*(2), 71–99.

Presidents' Alliance on Higher Education and Immigration (2023). 2022 Annual impact report. https://www.presidentsalliance.org/wp-content/uploads/2023/03/Pres-Alliance-Annual-Impact-Report-2022-web-spreads.pdf

Royster, J. J., & Kirsch, G. E. (2012). *Feminist rhetorical practice: New horizons for rhetoric, composition, and literacy studies.* Southern Illinois University Press.

Savage, G. (2013). Educating technical communication teachers: The origins, development, and present status of the course, "Teaching Technical Writing" at Illinois State University. *Communication & Language at Work, 2*(2), 3–19.

Savage, G. J. (1999). The process and prospects for professionalizing technical communication. *Journal of Technical Writing and Communication, 29*(4), 355–381. https://doi.org/10.2190/7GFX-A5PC-5P7R-9LHX

Savage, G. J., & Dura, L. (2023). Welcome to the first issue. *Technical Communication and Social Justice, 1*(1), 1.

Segal, J. Z. (1993). Patient compliance, the rhetoric of rhetoric, and the rhetoric of persuasion. *Rhetoric Society Quarterly, 20,* 90–102.

Skinnell, R. (2023). Why are there so many English PhDs, anyway? *College English, 86*(2), 111–135.

Skinner, D., & Franz, B. (2018). From patients to populations: Rhetorical considerations for a post-compliance medicine. *Rhetoric of Health & Medicine*, *1*(3), 239–268. https://doi.org/10.5744/rhm.2018.1013

Smith, B. (1985). Some home truths on the contemporary black feminist movement. *The Black Scholar*, *16*(2), 4–13.

Tamjeed, M., Tibdewal, V., Russell, M., McQuaid, M., Oh, T., & Shinohara, K. (2021, October). Understanding disability services toward improving graduate student support. In *Proceedings of the 23rd International ACM SIGACCESS Conference on Computers and Accessibility* (pp. 1–14). https://doi.org/10.1145/3441852.3471231

Tham, J. C. K. (2019). Feminist design thinking: A norm-creative approach to communication design. *Proceedings of the 37th ACM International Conference on the Design of Communication.*

Verhulsdonck, G., Howard, T., & Tham, J. (2021). Investigating the impact of design thinking, content strategy, and artificial intelligence: A "streams" approach for technical communication and user experience. *Journal of Technical Writing and Communication*, *51*(4), 468–492.

Walton, R., Moore, K., & Jones, N. (2019). *Technical communication after the social justice turn: Building coalitions for action*. Routledge.

Walwema, J., & Arzu Carmichael, F. (2021). "Are you authorized to work in the U.S.?" Investigating "inclusive" practices in rhetoric and technical communication job descriptions. *Technical Communication Quarterly*, *30*(2), 107–122. https://doi.org/10.1080/10572252.2020.1829072

8

WHAT WE CAME HERE FOR

Students Learning Local Civil Rights Rhetorics as Part of Kennesaw State University's Primary Source Initiative, the #ATLStudentMovement Project

Serenity Hill, Ahlan Filstrup, and Jeanne Beatrix Law

Introduction

In the evolving landscape of higher education, the integration of social justice in first-year college writing courses has become as imperative to many of us as the removal of diversity and inclusion initiatives has become increasingly ubiquitous. Our roles as educators often transcend semester-long academic instruction; we are tasked with preparing students for a world that is increasingly interconnected and diverse while simultaneously being polarized and disconnected. The authors of this chapter represent similar expertise in meeting this challenge, albeit at different life experience levels. We are an undergraduate Honors researcher (Ahlan); a graduate teaching fellow (Serenity); and a Professor of English (Jeanne). Collectively, we draw on our experiences to introduce first-year writing students to social justice through a local historiographic work, the #ATLStudentMovement Project. Our chapter explores the pedagogical effectiveness of and student feedback in deploying primary source writings with a focus on the roles Atlanta University Center students played in the civil rights movement of the 1960s. The use of primary sources in writing courses enables students to engage directly with the voices and experiences of people often marginalized in historical narratives. When students read documents such as *An Appeal for Human Rights* or listen to actual; testimonies from participants and leaders in the Atlanta Student Movement, they are immersing themselves in the lived experiences of individuals who fought for justice and equality. Students may also see themselves as the veterans of the 1960s Movement, as they too were around the same age

DOI: 10.4324/9781003469995-11

as our students when they staged political, economic, and legal challenges to Jim Crow segregation laws in the South. For our rhetorical and pedagogical inspirations, we drew on three scholars in the history of rhetoric, whose works spoke to us as we designed primary source assignments that would reach students in 2024.

Andrea Lunsford, in her well-known work on collaborative learning, emphasizes the importance of engaging students with diverse perspectives to foster a deeper understanding and appreciation of societal complexities (Lunsford, 2005). By incorporating such primary sources, we follow Lunsford's approach, facilitating an environment where students can critically engage with varied narratives, fostering empathy and a profound appreciation for diversity.

Jacqueline Jones Royster's call for inclusivity in the archival research process is particularly relevant in this context (Royster, 1996). Royster argues for the inclusion of underrepresented voices in academic discourse, asserting that their experiences and contributions are vital to a comprehensive understanding of history and society. By bringing these voices into the classroom, we not only honor Royster's vision but also provide students with a more nuanced understanding of social justice movements.

Lastly, Lynee Lewis Gaillet's work on historiography in composition underscores the importance of primary sources in developing critical thinking and writing skills in practical yet engaging ways (Gaillet, 2011). Gaillet posits that analyzing original documents from diverse groups cultivates a sense of historical empathy and enhances students' ability to contextualize contemporary social issues. In the context of ENGL 1102, this approach serves a dual purpose: it improves students' research and writing skills while simultaneously fostering a sense of personal connection to the broader human experience, particularly the struggles and achievements of those who have historically been silenced or overlooked, like the HBCU students and their allies who successfully fought to integrate public facilities and Atlanta Public schools over a two-year time period.

The justification for using primary sources from underrepresented groups in student research and writing is multifold. Firstly, it exposes students to a broader range of perspectives, challenging them to think critically and empathetically. Secondly, it encourages students to recognize the value of their own voices and contributions toward building a just and inclusive society. Finally, it aligns with the broader educational goal of fostering informed and engaged citizens, capable of contributing thoughtfully to societal discourse.

The integration of primary sources from underrepresented groups in teaching first-year college writing courses is not just an educational strategy; it is, indeed, a moral imperative. In technical writing contexts, specifically, efforts to include marginalized and underrepresented scholarly voices have led academia, particularly in the form of Cana Uluak Itchuaqiyaq's (2023) list of multiple marginalized and underrepresented scholars (MMU). The intentional acknowledgement and inclusion of underrepresented voices prepare students to become not only skilled writers but also empathetic, informed individuals committed to social justice and equity. As we designed our assignments for a research-focused ENGL 1102 course that could be scaled to multiple institutions, we thoughtfully considered the scholarship in this area, the advice of community members from the Movement, and our own knowledge of first-year students.

Research Question

As our team wrestled with design, learning modalities, and other intersections of student success, we settled on two key points: (1) the learning materials needed to be open educational resources (OER), and they needed to be primary sources that could be ingested by students in a format that appealed to them. Our research question then became: how will such a design enhance students' learning experiences, their sense of belonging, and pass rates in the course?

Review of Literature

A starting place to begin the correction of misrepresented history is in the classroom. In the education system, the hardships and work done by many Black Americans are not seen within the textbooks and lessons taught as a part of the curriculum. Sarah Bair (2020) states that educators and publishers believe that a more diverse and inclusive curriculum will foster more critical thinking and engagement while she believes the best place to start is with the contributions of women during the Civil Rights Movement to complicate and expand on the information already available (1). Her article discusses research conducted by Southern Poverty Law Center (SPLC) that describes how many high school students leave secondary education without a comprehensive understanding of the movement while her own research explains this could be a result of the four most used history textbooks having small portions dedicated to the movement and even less acknowledgement of the contributions of women (2–3). It is apparent that the information being introduced to future generations needs to be amended to introduce a well-rounded perspective of the period and promote diversity within classrooms. In doing so, students will be able to obtain a more accurate view of their history and better understand not only their own capabilities but also how the past influenced the present.

There is new literature produced that continues to address and support the recognition of southern Black women's significant involvement in the movement; however, there is still a gap within the literature that my research will fill. By not only looking at the Civil Rights Movement from a newer perspective, the perspective of not only women but Black women, but thinking beyond the most commonly discussed methods of support and protest such as participation in boycotts, marches, and sit-ins, the stories and lives of more women can be uncovered and shared. Barnet notes that "emerging scholarship and recent recollections are beginning to illustrate that, far from being apolitical or inconsequential, Black women were crucial to the Civil Rights Movement, that their personal experiences were unique as well as political, and that Black women's activism should be central to social movement scholarship" (p. 4). This research project is meant to do exactly that; not only discuss how crucial Black women's contributions were to the Civil Rights Movement, but to discuss how their personal experiences, as Black women and as writers, editors, and publishers, give a specific, unique look at the movement and the period. As a result of filling a gap in the present literature, I hope this research will influence both the educational curriculum and the publishing industry.

Methodology

The setting for our pedagogical students is a large public university, with a robust first-year writing program that is rooted in multimodal best practices with learning outcomes that support digital literacy and writing across the curriculum (Appendix A).

Student Demographics

Kennesaw State University (KSU) is the third-largest higher education institution in Georgia and in the top 50 largest universities nationwide. We welcome approximately 8,600 first-year writers each year, who join a community of more than 43,000 undergraduate and graduate scholars. We are situated in the North Atlanta Metro area on two campuses. More than a third of our students are first-generation; more than 50% of our students self-identify as non-white, with gender identification at about equal percentages. To teach this large number of students, our first-year writing program includes more than 100 faculty of all ranks and employment statuses including GTAs; we offer almost 800 sections of first-year writing each year. The learning outcomes for first-year writing include practicing the social aspects of the writing process and integrating sources with one's own voice to generate viable research projects. Statewide, public colleges and universities in Georgia also have general education learning outcomes. For first-year writing, these outcomes include communicating effectively for diverse audiences and purposes as well as source integration. Career-ready competencies include critical thinking and information literacy.

Together, the three researchers, professor, graduate teaching assistant, and undergraduate researcher, each contributed to the success of the project and we have our own reflections on the process of developing and executing the methodology. Below, we offer individual reflections, supplemented by our collective voices to articulate and present our curricular methodology. Ahlan and Serenity's reflections are contained here.

Ahlan's Reflection

As an undergraduate researcher, Ahlan offers a reflection on the process and project: The monumental impact of the Atlanta Student Movement cannot be understated. Hundreds of college students risked their education, freedoms, and lives in protests that were instrumental in the desegregation of Atlanta and the South as a whole. Their activism continues to inspire students of today, which we have consistently observed in a classroom setting. As educators in the Movement, we have found that an emphasis on community and open dialogue, coupled with primary source-based methodology, yields positive results in the overall learning experience. Student course feedback reinforces these observations. I will further reflect on my background as a learning assistant and undergraduate researcher throughout this section, presenting a narrative element to student course data.

In June of 2023, I was a learning assistant for a four-week Digital Community Engagement course that was taught by Dr. Jeanne Law. This upper-division class chiefly aimed to educate students on the Atlanta Student Movement's history through the lens of primary sources. Our class met twice a week for a total of eight in-person sessions. During the first class period, we provided students with physical and digital archival resources on the Movement. Materials included SOAR's repository of oral histories and the Atlanta Student Movement event timeline, as well as three digital folders of relevant photo and document scans. An informal verbal survey showed that none of the students had heard of the Atlanta Student Movement before attending the course. The archival materials functioned as a thorough, multimedia foundation for students' initial introduction to the Movement.

In-person class sessions typically consisted of open dialogue on the Movement. Classroom conversation was often rich with questions, proposals, personal anecdotes, and commentary on the Movement itself. We built a strong sense of community around these dialogues, which enhanced student interest and connection to the material. After their preliminary introduction to the Movement, students were instructed to create a digital "remix" on any topic that interested them. They were granted creative freedom in both the formatting and content of their final project. As a learning assistant, I advised students on final project ideas, development, and execution. Interestingly, several students gravitated toward the topic of women's activism. One person created an informative video about organizational leader Norma June Wilson Davis and her experiences at the Atlanta Prison Farm. The project used clips from Davis' oral history, interspersed with the student's own contextual commentary. It emphasized her intersectional identity as a Black female activist, describing the gender-based violence that was inflicted upon Davis and her peers. Another student created an informative infographic about various women of the Movement. Their project highlighted the lives of Gwendolyn Middlebrooks, Rosalyn Pope, Lydia Tucker Arnold, Ruby Dorris Smith-Robinson, Sarah Stephens, Norma June Wilson Davis, and Constance Curry. Each slide on the infographic included a quote from the activist, a short biography, and a description of her individual impact.

Overall, students displayed a strong desire to foreground the contributions of Black women, a marginalized group in civil rights discourse. This observation holds relevance in the context of teaching underrepresented activism. In his book *Living Black History*, Dr. Manning Marable (2006) discusses the power of reconstructing history through a "multidisciplinary methodology" (p. 18). "Historical narratives – the stories we teach about past events – become frameworks for understanding the past and for interpreting its meaning for our own time and in our individual lives" (p. 19), Dr. Marable (2006) states. Throughout the course, students learned about the importance of uplifting historically overlooked groups, with the Atlanta Student Movement as a central example. In turn, they chose to amplify the marginalized voices of women in their work. The result was stronger community engagement with underrepresented groups as a whole and a more accurate depiction of the Movement's diverse perspectives. Emphasis on social justice concepts in educational ventures may encourage students to explore underrepresented activism in their own communities.

Final course evaluations for the Digital Community Engagement course were generally strong. Although it was a limited pool of five responses, we received valuable feedback from those who did engage. 5/5 students responded "Strongly agree" to the following statement: "Overall, the content of this course contributed to my knowledge and intellectual skills." Elements such as access to primary sources, community-centered learning, and creative freedom were critical to this success. Written responses from the surveys showed that students particularly valued the flexibility aspect of our class structure. "Because students are able to research what they're interested in, the course content keeps them engaged with the class," said one person. Another student highlighted the importance of feeling heard by their instructor: "When students are speaking, she truly listen[s], then she gives valuable and thought-provoking feedback," the student wrote about Dr. Law. These sentiments reinforce the findings of several large-scale projects on student engagement, such as the 2012 report from the Student Retention and Success Programme. The UK-based report, completed with feedback from thousands of students over 22 higher education institutions, identified "active learning; group teaching; and enthusiastic and knowledgeable instructors" as significant variables in self-reported student engagement (Thomas, 2012, p. 37). The report further states that "some students may be less *intrinsically* interested in [a] subject, but engaging learning and teaching strategies can ignite interest and improve engagement" (p. 36).

Digital Community Engagement also received positive commentary on the resources that were provided as a foundation for student research. "I love that we have access to a plethora of supporting materials," stated one review. Teaching through primary sources, especially oral histories and discussions with Movement elders, provides a unique three-dimensional aspect to historical education. This element also draws attention to the locality and community of the activism, as students hear stories from activists who may have grown up in their own towns or held protests at well-known local establishments. Professor Libby Bischof (2015) remarks on this principle in a study, proposing that students who engage with local history become "public historians" in their own right (p. 549). Our class curricula continually emphasized the importance of primary sources as tools for learning. Through interacting with these sources, students developed a deeper understanding of why primary source preservation and dissemination is crucial for historical education.

Notably, several students expressed disbelief and frustration at the lack of mainstream education on local civil rights. "I lived in Atlanta for 18 years of my life and I didn't know any of this information," stated one student in the course evaluations. As we reflect on the current state of legal and social battles fought over DEI in the classroom, particularly in southern states such as Georgia, Florida, and Texas, we must recognize the necessity of making our voices heard. Creating open-source, accessible, engaging learning materials is one method we are continuing to develop in order to foreground historically accurate, inclusive social justice curricula. In light of the cultural climate surrounding DEI, we heavily prioritized the importance of historical education in general education courses. The

University System of Georgia mandates that all undergraduate students take 6 credit hours in writing-focused classes to attain their degree (University System of Georgia). At Kennesaw State, these required courses are designated English 1101 and 1102. The Atlanta Student Movement holds distinct potential as a learning tool in such courses. Its unique emphasis on local community engagement and historical social justice makes the Movement an ideal subject for first-year students to explore. Using Kennesaw State's first-year composition (FYC) guidelines (Appendix A), I designed twenty high-stakes assignment guidelines for first-year composition students. These assignments focus on the rhetorics of the Movement, align with Kennesaw State's English 1101 and 1102 standards, and encourage students to examine their own identities in relation to historical events. As an undergraduate student and educator in training, I considered my own experience in introductory writing classes, as well as the responses from Digital Community Engagement in my assignment design process.

Thoughts on Assignment Designs for ENGL 1102

Student engagement often presents unique difficulties due to each student's individual needs. In an upper-level course such as Digital Community Engagement, we as educators had the benefit of "looser" teaching guidelines. General education courses such as ENGL1101 and 1102 may present variable challenges in teaching social justice, due to the more structured nature of those first-year classes. Through my role as a learning assistant, I gained practical experience with students in the classroom. This exposure was greatly beneficial when tailoring lesson plans to English 1101 and 1102 standards.

Ultimately, I aimed to create assignments that engaged students and brought forward their individual strengths. I began by choosing sources that functioned as foundational resources for students. Archival materials were critical for this portion of lesson creation. In particular, I heavily incorporated the oral histories and the Atlanta Student Movement digital timeline into my assignments. All of the reference sources used in assignment guidelines are open education and zero cost. As stewards of the Movement's equity-focused principles, we strive for these educational materials to be accessible to all.

Furthering the theme of accessibility, assignments are designed using TILT formatting to offer students the best opportunity to succeed. TILT, which stands for "Transparency in Learning and Teaching," provides a standard outline to best emphasize clarity and motivate students in their learning (Winkelmes, n.d.). Each assignment sets forth a short "Purpose," "Task," and "Criteria for Success" description. Kennesaw State first-year composition standards mandate student completion of three major tasks for both English 1101 and 1102. "Narrative," "Analysis," and "Argumentative" are the required tasks for 1101, as students build their general writing skills at a college level. At the 1102 level, curricula focus more heavily on research-based tasks, mandating that students complete a "Research Review," "Research-Based Assignment" and "Research Project" over the course of the semester. Utilizing both TILT and institutional FYC standards, I built each assignment to reflect the appropriate structure of these courses.

We are currently in the early stages of implementing my lesson plans within the classroom. To ensure practicality, engagement, and overall student success, we chose a low-stakes pilot assignment that was incorporated into Serenity's English 1102 course. The assignment was initially a major research project, but we chose to adjust the guidelines for a smaller, simpler weekly assignment, keeping in mind the objective of garnering initial student feedback. Instructions asked students to choose one legal event along the Atlanta Student Movement timeline and research the law behind it. Located on Kennesaw State University's website, the timeline details important events related to the Movement, beginning with the Greensboro sit-in that inspired the Movement's creation. The purpose of the assignment was to provoke deeper thought on laws pertaining to segregation, while also strengthening student research and critical thinking skills.

TILT as a Best Practice

I designed the assignment based on Dr. Mary-Ann Winkelmes' TILT framework, which demonstrated success in "enhancing student success equitably" (Winkelmes, n.d.). The TILT framework was a foundational tool for us as we sought to engage students with the very language, structure, and rationale for writing assignments. As an undergraduate researcher, this part of the process was invaluable for me.

Serenity's Reflection

Classroom Context for ENGL 1102

As the teacher of record in a writing course and as a graduate teaching assistant, Serenity offers her reflections on executing the curricular methodology and scaling it to help prepare first-year writers to succeed in their other general education and upper-division major courses.

The key FYC Program Learning Objectives for English 1102 are to extend and reiterate the goals and objectives of English 1101, allow students to engage, analyze, and utilize various sources representing multiple per-spectives, and improve their writing and research skills by entering an on-going conversation. Dylan B. Dryer (2016) emphasizes writing, and by proxy research, are not innate skills; therefore, they must be developed, but as a general education course, instructors must consider how to cre-ate an environment in which students are invested in their development (p. 29). As previously mentioned, educators are challenged with prepar-ing students to address a society that is increasingly interconnected and diverse while simultaneously being polarized and disconnected. In my courses, in addition to the institution's objectives, I help students engage in research that introduces them to the world beyond the classroom, inves-tigates their own communities, and explores how to connect their identity to their writing. To combat divisiveness in a diverse classroom, I begin English 1102 by introducing, defining, and discussing identities, communi-ties, and how they not only connect us but can also be implemented into our discussions of writing and research. In short, I set up the framework for our scholarly discussions, allowing for student input but also setting up parameters to ensure an inclusive space. Keeping community and identity in mind, I theme my course around true crime, which is meant to allow students to explore what major events affect their community, why they are interested in said events, and how these events are connected to the world around them.

As the English 1102 course objectives focus on the use of research to improve student writing, I chose to use the assignment guidelines from OER major assignment developed by Ahlan for a 1102 course to create a low-stakes, in-class group assignment best fit to generate deeper under-standings about the impact of writing as a response to significant events that influence our connected communities. The goal in choosing to re-work this assignment into a scaffolded activity was to create an exercise that not only aligned with the learning outcomes of the course and was pedagogically effective but also allowed students to uncover the lasting impression social justice has on current cultural moments in the United States. During this time in the semester, students were working toward creating their annotated bibliographies for their final research essays, choosing, investigating, and analyzing the credibility and significance of their sources.

Analysis and Discussion

As the instructor of record for the 1102 course in which the assignment was tested, Serenity provides context and key instructional choices for the assignment designed by Ahlan.

Kevin Roozen (2016) asserts that writing is a social, collaborative, and rhetorical act (p. 18) while Graff and Birkenstein (2021) highlight writing as a response to a larger conversation (p. 6). Larger conversations and responses can be created due to everyday problems or event-based problems (Nicotra, 2023, p. 112). Keeping these notions and the TILT method in mind, I chose to make the in-class scaffolding activity, based on the OER major assignment "Timeline | Research", a group activity using "An Appeal for Human Rights", a pivotal moment in the Atlanta Student Movement timeline, as the reference material. "An Appeal for Human Rights" was the perfect text for this course because it is an excellent example of writing and research as a response to everyday problems that became an event-based problem as it directly addresses everyday issues as a result of the Jim Crow era and reflects how our identities influence our writing and connects many communities.

Referencing the Classroom Context for English 1102, a goal of my class is to introduce society beyond the education system and support students' inquiry into how identity, writing, and research intersect. I began the activity by introducing *An Appeal for Human Rights*, providing contexts for students such as it was published in 1960 during the Civil Rights Movement, written by students in their region for the betterment of all mankind, not just black and white people, so they not only saw the usefulness of text as entering a conversation, but the impact social justice in their community then and now. The activity began with students reading "An Appeal for Human Rights" and the class discussed the connection between this text, their identities, and writing as a response. Following this discussion, students were placed into groups to research events, laws, or policies related to one of the seven sections of the *Appeal for Human Rights* their group felt most connected to.

Within these groups students were asked to:

1 Create a research question using the keywords: equity, unequal, issues, policies, laws, or events.
2 Find a source to help answer their research question.
3 Answer the following for their source:

Who: Who wrote it?
What: What was said? What new information did you find?
When: When was it published? Relevance?
Where: Where was it published? Is this credible?
Why: Why is this information significant to your research question?
How: How does this connect to the "Appeal for Human Rights"?

4 Discuss this event, law, or policy's impact/effect on you or your community as a group.

After students completed their work within their groups, students were asked to share their group's findings to class, allowing everyone to see the relatedness of their communities and the impact social justice work has on people with diverse backgrounds. As a "ticket out of the door" students were asked to submit anonymous feedback about their experience with this lesson as it pertained to their understanding of the research process, writing, social justice, and their engagement with the in-class activity. The feedback questionnaire was comprised of four close-ended questions and one close-ended question:

1 This assignment was helpful to me in understanding the beginning of the research process.
2 This assignment was useful to me in my understanding of writing.
3 I enjoyed this assignment.
4 This assignment encouraged me to think deeper about the impact social justice has in my life.
5 My thoughts on how the assignment worked or didn't work for me.

We analyze trends and findings here, with the complete data set offered in Appendix B.

Discussion of Results

Empowering students to be stakeholders in their own learning and implementing a democratic pedagogy are cornerstones for all of our teaching philosophes These principles were applied in the design of this activity, allowing students to select their groups, choose their topics from the source material, compose their research question, and evaluate and assess the information they found in response to their research questions. Individual student engagement with this activity could be impacted by many different factors such as having limited time to complete the assignment as it was done in one fifty-minute session and working in groups rather than individually. Still, this mixed method study prompted student responses to be dictated by their feelings and the results reflect an overwhelmingly positive pilot.

Question one assessed this exercise's helpfulness in understanding the beginning of the research process. With ninety-seven percent of the students agreeing they found the activity beneficial in aiding their understanding of writing, and three percent disagreeing, it was successful. A goal while executing this assignment was to make sure supplemental assignments designed based on the OER materials created by Ahlan would align with FYC Program Learning Objectives for English 1102, which is to improve student comprehension of the writing and research process.

Research begins with inquiry and inquiry starts with a question. By constructing a low-stakes activity that begins with students crafting a research question and then analyzing the information they found in response to their questions meets the goals of FYC. As mentioned before, the FYC Program Learning Objectives not only focused on enhancing student understanding of the research process but also on improving and expanding their skills and knowledge of the writing process. Student responses indicate the design of this assignment assisted in their understanding of writing as a tool as ninety-one percent of students agreed it aided them and the remaining nine percent had no opinion. *An Appeal for Human Rights* allowed students to see the relationship between writing, research, and rhetoric, in addition to being an example of writing and research as a rhetorical tool to respond to problems, they were also able to use it to enter an ongoing conversation about social justice and its effect on them. Survey Question 3 gauges how much students enjoyed the assignment. As shown in Figure 8.1, there were three possible answers and percentages for each that will help us get a sense of students' attitudes toward the assignment design, presentation, and organization.

Question three directly reflects student engagement with the assignment as it asks how well they enjoyed participating in the activity. Figure 8.1 presents the student responses as thirty percent expressed they enjoyed the

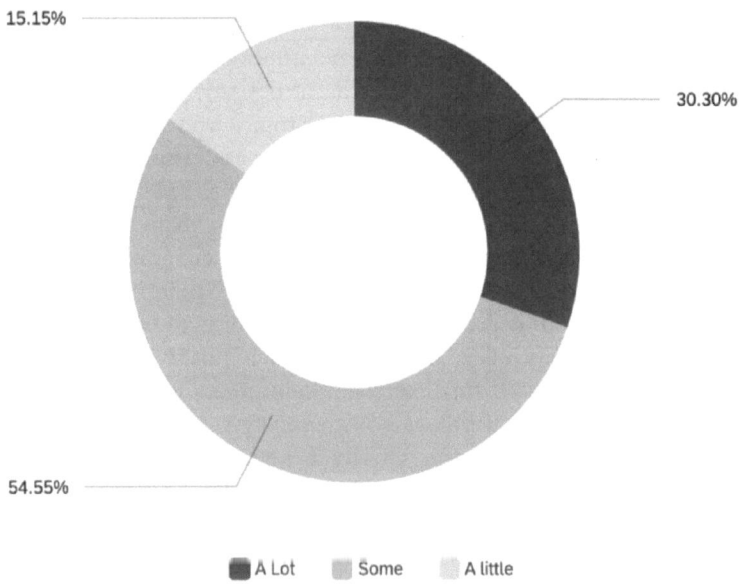

FIGURE 8.1   Student responses to Question 3: "I enjoyed this assignment."

assignment a lot while fifty-five percent of students enjoyed it some and fifteen percent enjoyed it a little. If we collapse the statements some and a little, nearly seventy percent of students were less entertained with the activity as opposed to the remaining thirty percent who were very fond of the assignment. Despite the lack of enthusiasm for the assignment, students did have the option to select if they did not enjoy the assignment at all and no respondents did. While the initial responses gave us a snapshot of student attitudes, we did not measure variables that could influence these degrees, such as likes or dislikes of group work, group dynamics, and in-class assignment length. As we continue testing this assignment, we will do a deeper dive into the "why" that the preliminary answers to Question three indicate.

In analyzing the efficacy of the assignment, we also wanted to assess if it encouraged students to connect the maxims of *An Appeal for Human Rights* to their own lives. Figure 8.2 shows preliminary results of this assessment. Kennesaw State University has a unique demographic that is reflected within Serenity's diverse classroom. Despite a student's background or identity, social justice has influenced their life. Figure 8.2 depicts the result of question four which asks if students felt encouraged to think deeper about the impact of social justice in their lives. Sixty percent

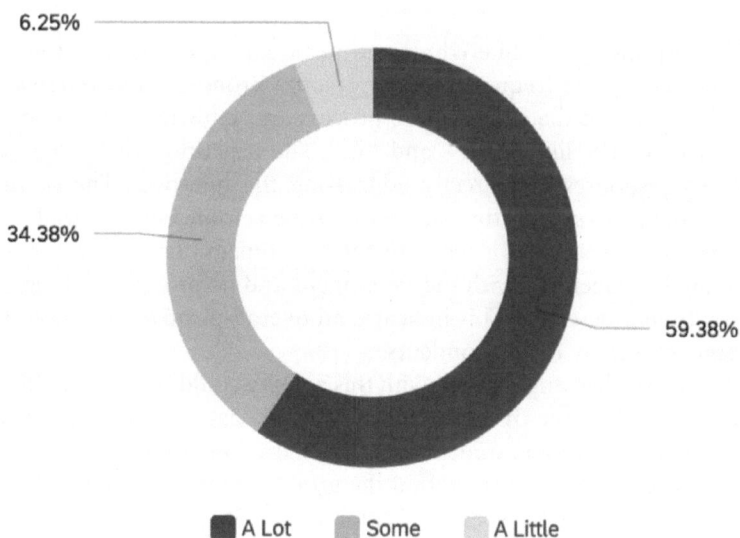

FIGURE 8.2   Student responses to Question 4: "This Assignment Encouraged me to Think Deeper about the Impact Social Justice has in my Life."

of students expressed that the activity encouraged them a lot to think about the impact of social justice while thirty-four percent said they felt somewhat encouraged and six percent were a little encouraged. Like the responses to question three, students did have the option to select if they did not feel encouraged by the assignment at all to think deeper about social justice and no respondent did. Collapsing the some and a little response again, the data shows sixty percent of students were able to critically think about the influence of social justice in their life while forty percent of students considered the impact but did not greatly expand their thinking on the subject.

Question five was an open-ended question that asked for students' thoughts on how the assignment worked or didn't work for them. We assembled a sentiment analysis for the written responses. The analysis yielded 14 positive and 14 neutral responses. There were no negative responses indicated. Key insights include:

1   Overall sentiment of the responses is slightly positive, as indicated by the average polarity.
2   Half of responses are positive; the other half are neutral. No explicitly negative responses.
3   Average subjectivity is moderate, suggesting a balance between personal opinions and objective statements in the responses.

The responses are centered around personal experiences and impacts, as indicated by the frequent use of personal pronouns and references to the self. Focus on the assignment's process and influence on respondents, as seen in words like "how" and "it." The repetition of "assignment" confirms responses are directly addressing the question. The sentiment and textual analysis together suggest that the respondents generally had a positive or neutral experience with the assignment. The responses reflect a personal engagement with the assignment and its process, with no negative sentiment detected. This indicates an overall positive reception of the assignment among the respondents.

Overall, student engagement with this activity could be a direct reflection of the group dynamic that was in place for this assignment. During group work, many times some students who are more interested in a topic take a leadership role while others within the group decide to do only what they are instructed within the group. Additionally, due to groups, students had to collaborate on the topic they chose to investigate from the reference material; therefore, if a student was not attached to the topic, their interest in the assignment may have faltered. Frequently, this results in the students who take a backseat approach to the group work being less engaged with the assignment and the objectives it is meant to meet. Despite the possible influential factors such as the assignment being in class, group work, and time limits, students still found the assignment engaging and were able to think critically and unveil the lasting impact social justice work has on their communities. Crafting work that supports collaborative learning and highlights the importance of sharing diverse perspectives, through archival primary sources, for students to engage with, motivates critical thinking and writing skills while aligning with FYC Program Learning Objectives in practical yet innovative ways.

Design and Learning Activities

A key aim of Kennesaw State's first-year writing program is to prepare students for successful writing not only in their general education courses but also in their majors and post-graduate careers. Part of meeting this aim requires innovative instructors who can envision the application of learning activities in first-year writing courses that can be scaled to across-the-curriculum student success. We believe that the curricular methods and content we developed and iterated in an ENGL 1102 course can be scaled and extended to meet learning outcomes and program goals of other types of writing courses and learning environments. The three primary connections we see to how our design can be applied across learning contexts are:

1 We designed the social justice component of this project to elicit students' self-reflections in being able to place themselves within the context of a larger cultural shift. We accomplished this goal by using a local, historical connection to the Civil Rights Movement of the 1960s. Students provided us with valuable feedback on this part of our design. Almost no students knew about the role Atlanta college students played in the Movement. We advocate for instructors in other writing course contexts to seek out local civil and human rights examples to assist their own students in placing themselves within these contexts, that they often see and pass by every day.

2 The learning activity itself allowed students to learn writing and research skills that expand beyond the first-year writing classroom while also creating a space for students to investigate issues that affect their communities. This low-stakes, in-class assignment gives students the opportunity to learn and practice the beginning stages of research using archival material to collaboratively create research questions, use keywords, and located credible sources. The feedback provided by students shows that the activity improved student understanding of the research process by introducing archival and library tools and promoting

qualitative analysis and critical thinking which is necessary across many curriculums and disciplines. We encourage educators to integrate learning activities that teach new skills and allow students to explore their communities and identities from different perspectives.

3 Just as we teach our students to engage in a larger conversation, the design and learning outcomes of this assignment exist within a larger curricular context. By discussing landmark documents such as *An Appeal for Human Rights* and using it to support metacognition, we are integrating diversity, equity, and inclusion into writing environments as this assignment is a testament to the benefit of an inclusive teaching pedagogy. The goal is to create a respectful, diverse, and socially aware community. By incorporating DEI principles, writing courses can build an environment that values diverse perspectives, promotes self-awareness, and provides equitable opportunities for all students to develop their writing skills effectively.

Implications for Future Research

Our initial data from the assignment beta test shows positive correlations between the assignment design and student attitudes toward its social justice goals. As researchers, we believe that the outcomes of this study illustrate significance. As instructors who practice social justice as a pedagogical grounding, we believe that we successfully wove the themes of social justice through a low-stakes assignment that will lead students toward deeper and abiding understandings of how historical human rights struggles are, indeed, still their own as well. This assignment was aligned to KSU's first-year writing Program Guidelines and to the University System of Georgia's General Education Refresh (University of Georgia). Certainly, part of the assignment's significance and value comes from alignment to curricula. What we mean by this is that in institutional learning environments where social justice themes may come under attack, aligning that work to best practices and standards allows us to provide accurate, consistent outcomes while generating critical thinking and information literacy learning that are part of most first-year writing programs.

We invite scholars from across disciplines to test this assignment, make it their own, and engage with us about their own results. Studies such as ours give teacher-scholars the opportunity to embed critical, cultural moments into our first-year writing courses and allow students to re-engage with historical civil rights moments in ways that cultivate meaning for them. These types of classroom engagement also serve to sustain deep thinking and critical literacies that help us sustain the vital stories and experiences that will help students of today become the empathetic leaders of tomorrow.

References

Bair, S. (2020). The American civil rights movement reconsidered: Teaching the role of women. *The Social Studies, 111*(4), 1–9.

Bischof, L. (2015). The lens of the local: Teaching an appreciation of the past through the exploration of local sites, landmarks, and hidden histories. *The History Teacher, 48*(3), 529–559.

Dryer, D. B. (2016). Writing is not natural. In L. Adler-Kassner, & E. Wardle (Eds.), *Naming what we know: Threshold concepts of writing studies* (pp. 27–29). Utah State University Press.

Gaillet, L. L. (2011). Historiography in the writing classroom: A project in archival research methods. *Teaching English in the Two-Year College, 38*(3), 244–255.

Graff, G., & Birkenstein, C. (2021). *They say/I say: The moves that matter in academic writing.* W. W. Norton & Company.

Itchuaqiyaq, C. U. (2023). MMU scholar list. https://www.itchuaqiyaq.com/mmu-scholar-list

Lunsford, A. (2005). Collaboration, control, and the idea of a writing center. *The Writing Center Journal, 15*(1), 3–10.

Marable, M. (2006). *Living Black history: How reimagining the African-American past can remake America's racial future.* Civitas Books.

Nicotra, J. (2023). *Becoming rhetorical: Analyzing and composing in a multimedia world.* Cengage Learning.

Roozen, K. (2016). Writing is a social and rhetorical activity. In L. Adler-Kassner, & E. Wardle (Eds.), *Naming what we know: Threshold concepts of writing studies* (pp. 17–19). Utah State University Press.

Royster, J. J. (1996). When the first voice you hear is not your own. *College Composition and Communication, 47*(1), 29–40.

Thomas, L. (2012). Building student engagement and belonging in higher education at a time of change. Student Retention and Success Programme Report.

University of Georgia. (2023) 3.3.1 core curriculum: Core IMPACTS. Board of Regents Policy Manual.

Winkelmes, M.-A. (n.d.). TILT higher ed examples and resources. *TILT Higher Ed.* https://www.tilthighered.com/resources

APPENDIX A

Kennesaw State University's first-year composition guidelines

FYC Program Pathways: Alignment and Consistency for 1101 and 1102

Overview

The following major assignments for ENGL 1101 and 1102 will help generate consistent language and engender alignment across the FYC program at KSU. In creating these assignment structures, FYC faculty working groups have considered multiple stakeholders, including: students, colleagues, and administration (in this case USG) – in that order. Moreover, the working groups created these assignment sequences in consultation with FYC faculty at KSU and after careful consideration of faculty syllabi and best practices comparator and aspirational institutions. Adopting these frameworks in our classrooms will ensure that students have consistent and transferrable learning experiences in FYC, that our colleagues have creative license to innovate in their classes, and that administration has necessary metrics against which we can map student learning.

For our work, we delineate differences between **types** of assignment and actual assignments themselves. The **type** of assignment should be overarching and not include specific instructions or rubrics. For example, an "analysis" (see 1101 section below) is a "type" or "genre" of writing. An instructor's assignment that aligns to that type might be framed as a multimodal text, may be an essay that asks a student to consider how an author approaches purpose/audience/style/context, or it may even be a deep dive into a social issue that has been written about across publication contexts.

Whatever the assignment looks like in form and function, as long as it meets the criteria to be considered an analysis, then it is aligned appropriately.

As we develop a programmatic future that aligns with trends in the field of FYC, comparator and aspirational institutions, and USG expectations, it's important to remember that this is a "living document." Program growth requires that we retain flexibility to revise this document as trends in the field change, as we situate ourselves uniquely within the USG, and as we consider KSU as a model for embracing best practice trends in the field. **For now, these assignment types continue to be in-place for AY 2022-2023.**

Guiding Principles for Use of These Assignments in the Classroom

1 We will have a genre/type based assignment sequence in FYC going forward so that our students, faculty, and other stakeholders can expect consistency of learning as well as local, sustainable, and measurable data that shows student success.
2 We will have a syllabus style/template that gives faculty the freedom to determine percentages for major assignments, the actual assignments they can align to the required **types** for each gateway course, their texts of choice selected from the current FYC approved list, and scaffolding/pacing of their course.
3 These are major assignment types only; three (3) major assignments per course.[1]
4 Low-stakes and scaffolded work, as well as day-to-day lesson plans are left up to individual instructors.
5 Assignments aligned with these types need not be traditional papers (though they very well may be). While writing will be an important component of any major assignment, the final product may take the form of alternative media or be multimodal in accordance with the instructor's course structure and pedagogy.

Brief Overview of Course Assignments

Every section of our FYC courses will align with the relevant major assignment sequences below. More details on each are provided in the following section. Sample assignments and syllabi from colleagues, TILTed resources, and examples of rubrics are located in the FYC D2L portal. **Please note: Effective Summer 2022, the first two assignment types for ENGL 1102 can be assigned in the order below or reverse order, depending on an instructor's pedagogical strategies.**

Engl 1101

1 Narrative/Reflective Assignment
2 Analysis Assignment
3 Argument Assignment

Engl 1102

1 Research Review/Annotated Bibliography
2 "First Look" Research-based Assignment
3 Research Project

Detailed Overview of Course Structures, Learning Outcomes, and Assignment Descriptions

Engl 1101

English 1101 Course Description

English 1101 focuses on skills required for effective writing in a variety of contexts, with emphasis on exposition, analysis, and argumentation. Also includes introductory use of a variety of research skills.

English 1101 Course Outcomes

Upon completion of English 1101, students will be able to...

1 Practice writing in situations where print and/or electronic texts are used, examining why and how people choose to write using different technologies.
2 Interpret the explicit and implicit arguments of multiple styles of writing from diverse perspectives.
3 Practice the social aspects of the writing process by critiquing your own work and the work of your colleagues.
4 Analyze how style, audience, social context, and purpose shape your writing in electronic and print spaces.
5 Craft diverse types of texts to extend your thinking and writerly voice across styles, audiences, and purposes.

Major Aligned Assignments

1 **Narrative/Reflective Assignment:** a narrative assignment asks students to write through a *story* or a *history* (or both). This could take the form of a reflection. Chronology is a synonym here as well.

Sample assignments include but are not limited to:

- Personal narrative
- Collage/multi-genre narrative
- Literacy narrative/digital literacy narrative
- "This I Believe" assignment
- Transition narrative (into college or another important transition)
- Family history
- History of a profession or work narrative
- Reflective portfolio cover letter

2 **Analysis Assignment:** an analysis assignment asks students to break some object of study into its component parts and examine those parts carefully to come to a better understanding of the whole.
    Sample assignments include but are not limited to:

- Rhetorical analysis
- Genre analysis/comparative genre analysis
- Audience analysis/kairotic analysis/rhetorical situation analysis
- Stakeholder analysis
- Text in context analysis
- Pattern + interpretation
- Conceptual lens/interpretive lens analysis

3 **Argument Assignment:** an argument assignment includes persuasion as an explicit goal. Persuasion can be broadly conceived; this assignment need not involve taking a stand on a controversial issue (though it may).
    Sample assignments include but are not limited to:

- Persuasive assignment
- A specific argument method: Rogerian, Toulmin, etc.
- Causal argument
- Definition argument
- Op-ed (or another "public" argumentative genre)
- Joining the conversation or they say/I say essay

Engl 1102

English 1102 Course Description

English 1102 focuses on developing writing skills beyond the levels of proficiency required by ENGL 1101. Emphasizes interpretation and evaluation and advanced research methods.

English 1102 Course Outcomes

Upon completion of English 1102, students will be able to...

1 Locate print and digital sources that represent multiple perspectives.
2 Analyze sources by critically reading, annotating, engaging, comparing, and drawing implications.
3 Practice working through the writing process, including brainstorming, drafting, peer review, revision, and publication.
4 Compose a rhetorically-situated, researched text that enters an ongoing conversation, integrating relevant sources.

Major Aligned Assignments[2]

1 **Research Review:** in this assignment type, students present key insights gathered from the research they have been conducting as they work toward developing their research project's thesis. In producing a research review, students use databases and other university (and other) resources to find research materials; assess the quality of that research in relation to a larger, ongoing research project; and demonstrate appropriate academic documentation style.
   Sample assignments include but are not limited to:

   • Literature Review
   • Annotated Bibliography
   • Journal of notes/note cards (submitted for review)

2 **"First Look" Research-based Assignment:** students present their projects' topic, preliminary research, tentative thesis and/or potential argumentative points at an intermediate stage of the research-project process for feedback from peers and/or the instructor.
   Sample assignments include but are not limited to:

   • Outline
   • Précis
   • Proposal
   • Rough Draft (submitted for a grade)

3 **Research Project:** the research project represents the culmination of the recursive practices of the course. Students will present a polished product of their work that illustrates the development of the project from the aforementioned stages. Consequently, the project should include a properly-documented, carefully-developed argument that makes use of research.

Sample assignments include but are not limited to:

- Research Paper
- Researched Essay
- Multimedia Project
- "Ted Talk"

Notes

1 This is a minimum number of assignments. Faculty may choose to offer more on their own.
2 Please note: ENGL 1102 is not a literature-based course. We do not teach literary research in this course. Also note: effective Summer 2022, assignments 1 and 2 may be taught in reverse order.

APPENDIX B

Full survey report

Q1 - This assignment was helpful to me in understanding the beginning of the research process

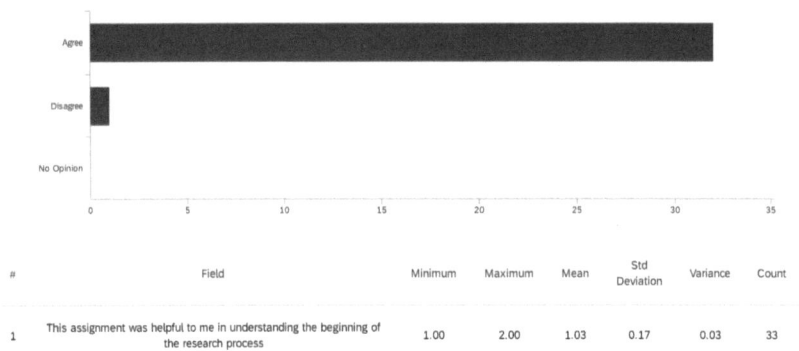

#	Field	Minimum	Maximum	Mean	Std Deviation	Variance	Count
1	This assignment was helpful to me in understanding the beginning of the research process	1.00	2.00	1.03	0.17	0.03	33

#	Field	Choice Count	
1	Agree	96.97%	32
2	Disagree	3.03%	1
3	No Opinion	0.00%	0
			33

Q2 - This assignment was useful to me in my understanding of writing

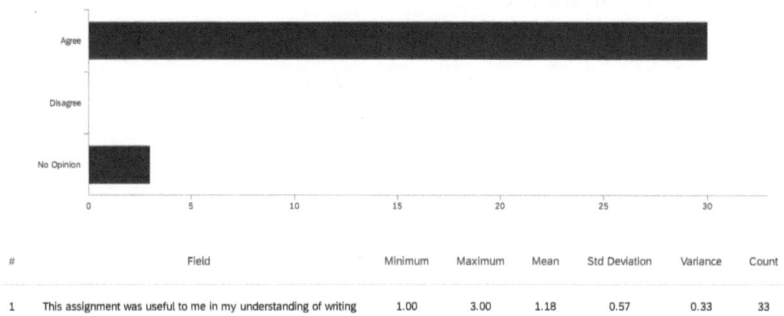

#	Field	Minimum	Maximum	Mean	Std Deviation	Variance	Count
1	This assignment was useful to me in my understanding of writing	1.00	3.00	1.18	0.57	0.33	33

#	Field		Choice Count
1	Agree	90.91%	30
2	Disagree	0.00%	0
3	No Opinion	9.09%	3
			33

Q3 - I enjoyed this assignment

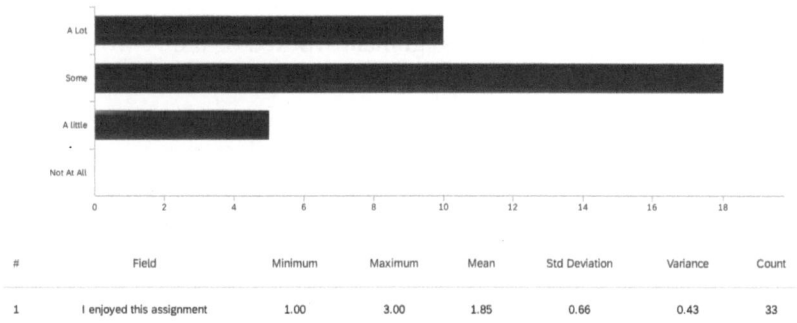

#	Field	Minimum	Maximum	Mean	Std Deviation	Variance	Count
1	I enjoyed this assignment	1.00	3.00	1.85	0.66	0.43	33

#	Field		Choice Count
1	A Lot	30.30%	10
2	Some	54.55%	18
3	A little	15.15%	5
4	Not At All	0.00%	0
			33

Q4 - This assignment encouraged me to think deeper about the impact social justice has in my life

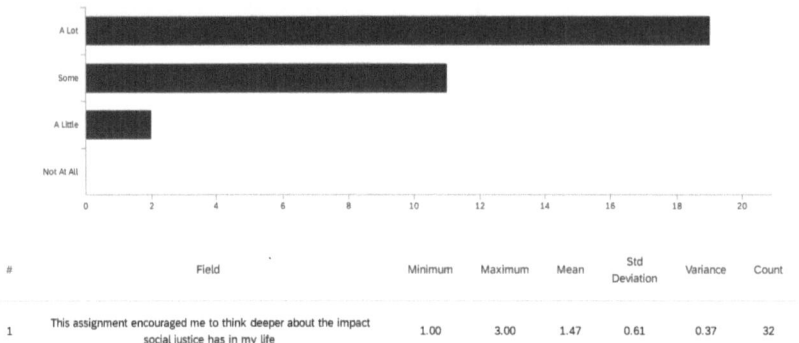

#	Field	Minimum	Maximum	Mean	Std Deviation	Variance	Count
1	This assignment encouraged me to think deeper about the impact social justice has in my life	1.00	3.00	1.47	0.61	0.37	32

#	Field		Choice Count
1	A Lot	59.38%	19
2	Some	34.38%	11
3	A Little	6.25%	2
4	Not At All	0.00%	0
			32

Q5 - My thoughts on how the assignment worked or didn't work for me

My thoughts on how the assignment worked or didn't work for me

It allowed me to create a research question that allowed to find resources that allowed me to answer and explain the research question

It gave me a deeper understanding about human rights.

Yes it worked because it showed me the fundamentals of research and writing properly.

It helped me understand the research process a little more

It forced me to be more opinionated and think of an answer quicker cause i was woking with people

It gave me a better understanding on how I should do my future research.

This assignment helped me understand how to go about researching a topic and writing about it.

I think it worked for me to understand the question better in order to get a better understanding.

It helps get students thinking about how to scratch the surface when it comes to the research process

My thoughts on how the assignment worked or didn't work for me

I worked for me because it was simple and easy to understand.

This assignment was very informative and helped me engage in learning much better

n/a

This was a good assignment so it worked.

I liked how you gave a set article and got us to find a topic related to it i think that helped understand how to figure out a topic based on something else to start a research essay

My thoughts on how the assignment worked for me was very good and exciting because I learned something new.

I honestly like these types of these assignments, it helps me in ways I didn't think were possible.

Opened my eyes to the differences blacks face amongst PWIs

This assignment helped with being able to search for information to answer questions.

It helped me process these topics on a different level. I was able to fully dig deeper into the standards

It worked because it was involving and i got a deeper sense of what were learning and applying this.

it was a great experience

It helped me get started on the thinking process needed for the project.

made me think, really cool, lol thanks.

9

"NOTHING ABOUT US WITHOUT US"

Post-Exertional Malaise (PEM) and the Challenge of Designing Activism for People with ME and Long Covid

Jennifer Nish

On a warm, sunny day in September 2022, I was in Washington D.C. at a protest focused on drawing press attention to complex chronic diseases such as myalgic encephalomyelitis (ME) and Long Covid. ME and Long Covid are both complex chronic diseases that can be caused by a virus. There are documented cases of ME going back to 1926, including outbreaks associated with other viruses such as polio. Experts have estimated that 10–30% of people who get COVID-19 go on to develop Long Covid, and nearly 50% of people with Long Covid meet diagnostic criteria for ME (Davis et al., 2023; Mancini et al., 2021). Although ME existed prior to the COVID-19 pandemic, many organizations, researchers, and clinicians with expertise in ME now include both ME and Long Covid as part of their focus. While some public health experts have claimed that Long Covid and ME are the same (see Phillips & Williams, 2023), this opinion is not necessarily widespread, so I use both terms in this chapter. Both diseases are debilitating and there are currently no FDA-approved treatments for either disease (Figure 9.1).

In this chapter, I use a combination of narrative and a design case study to show how activist engagement (Nish, 2022) educates participants about putting political principles into action. I argue that disability is a source of embodied knowledge from which designers and activists can learn, regardless of their own experience with or personal relationship to disability.

DOI: 10.4324/9781003469995-12

FIGURE 9.1   Photograph (#MEAction, n.d.) of me standing in a group of other ME and Long Covid activists, holding a sign that reads "We Deserve Better."

Rhetoric, Technical and Professional Communication, and Design (In)Justice

Scholars in technical and professional communication (TPC) have used design as a way of approaching problems, and particularly complex, or "wicked" problems (Tham, 2021). In disability studies, discussions of design most commonly refer to universal design, which is a framework for addressing the needs of a range of people in the design of products, resources, events, spaces, and other human creations. Sometimes universal design places particular emphasis on disabled people, but as Aimi Hamraie (2017) shows in *Building Access,* universal design has an inconsistent history of prioritizing and/or emphasizing disabled users' needs.

ME and Long Covid activism is an example of a particularly complex problem that shows how design can be used to facilitate inclusion and access for disabled people. Designing a protest for people with ME and Long Covid requires attention to embodied capacities, which include significant differences in function between participants.

An interconnected set of rhetorical problems shapes Long Covid and ME activism. Drawing on Jay Dolmage's (2014) definition of rhetoric as "the circulation of power through communication," some of the rhetorical problems that people with Long Covid and ME face are

- lack of public awareness about what these diseases are;
- inadequate data about who has these diseases;
- lack of knowledge among primary care doctors and most other clinicians about how to treat symptoms and support patients;
- scarcity of specialists who focus on these diseases;
- poor fit between complex, multi-system chronic diseases of unknown etiology with Western medicine's and U.S. healthcare's categorization of disease and treatment through specific bodily parts and systems; and
- lack of awareness about the disproportionate impact that ME and Long Covid have on marginalized populations (such as Black, Latinx, and trans people), in which these diseases are more common yet underdiagnosed.

Scholars have characterized the experiences of people with ME and Long Covid as a form of epistemic injustice (Callard & Perego, 2021). Miranda Fricker (2007) describes epistemic injustice in two forms: testimonial and hermeneutic. In testimonial injustice, marginalized and oppressed people speak about their experiences but audience members from the dominant group do not hear or understand them. This failure occurs because the audience holds biases about the speaker's identity that undermine the speaker's credibility. In the second type of epistemic injustice— hermeneutical injustice—groups of marginalized and oppressed people are unable to contribute to collective knowledge because dominant ways of understanding and knowing do not fit their experiences. This injustice can interfere with marginalized and oppressed groups' self-understanding as well as the dominant culture(s)'s ability to hear and incorporate the oppressed groups' knowledge.

Both types of injustice apply to people with ME and Long Covid. One of the central problems that people with ME and Long Covid face is testimonial injustice. Researchers, public health officials, doctors, and others have historically dismissed people with ME in a variety of ways, such as claiming that people with ME are overreacting or that their physical symptoms are caused by mental or psychological factors (Brea, 2017; O'Rourke, 2023). Journalists and experts have linked this history to data showing that ME is much more common in women and women with ME tend to experience more symptoms than men with ME (Devlin, 2023; Ford, 2018; Hsu, 2023; Yong, 2020b). This history of dismissal was repeated as people with the earliest cases of Long Covid were dismissed by doctors, public health officials, and the people around them (Yong, 2020b). Even now, access to knowledge about Long Covid and techniques for managing symptoms varies significantly according to class, race, geographic location, and other factors that shape health and healthcare access. People with Long Covid and ME face an uphill battle as they attempt to raise awareness about these diseases and the need for research and treatment. Because phenomena like post-exertional malaise contradict both cultural and medical truisms (e.g., that exercise will improve health), ME and Long Covid are difficult to integrate into existing schemas. However, people with ME and Long Covid are also designing innovative activist campaigns to resist these injustices.

#MillionsMissing 2022

Back to the day of the protest: under the hot afternoon sun, a group of about 50 people moved in a small circle on the sidewalk in front of the White House. I was holding a sign that read "We Deserve Better" and chanting with other participants. Our group moved slowly back and forth in front of the quintessential view of the White House and its large grass lawn. Our protest demanded that President Biden declare Long Covid and ME a national emergency, which would provide a mechanism for infusing funding into research on desperately needed treatments for Long Covid and ME.

Organizers chose the White House location for a few reasons, including its recognizability as a symbol of the U.S. government and its power. Beyond the message to President Biden, organizers' goal was to draw media attention to the protest. Of course, we were not surprised when Biden did not declare a national emergency in response to our work. The protest received press coverage in major media outlets such as *New York Times*, *NPR, Time,* and several medical news outlets (Tillman, 2022). Several prominent journalists and disability activists also posted or shared material about the protest on social media. The press regularly used the syndicated photos of the #MillionsMissing protests to illustrate Long COVID articles over the following year.

As I have written elsewhere, asking whether a protest achieved its central objective is not necessarily the best way to evaluate its effects.[1] Whether or not a protest results in participants' ideal outcome or response, participating in protest affects participants (and possibly also observers) in various ways: demonstrating or reinforcing personal commitment to an issue; showing public audiences and/or political representatives how much collective support there is for a movement or issue; building community among participants; building and/or maintaining enthusiasm and energy; demonstrating political principles through action; and communicating with other protesters and onlookers through signs, chants, t-shirts, and other forms of protest rhetoric.

Our protest initially followed guidelines set out in our permit from the National Parks Service, but we also planned to engage in nonviolent civil disobedience to draw attention to our cause. Our protest permit specified that we needed to keep moving when protesting on the sidewalk in front of the White House. At a designated time during the protest, a small subgroup of protesters laid down on the sidewalk in front of the White House in violation of the permit. The goal of this action was to get arrested, drawing inspiration from the civil disobedience of HIV/AIDS activists.[2] The goal of getting arrested was to draw media attention to the protest and to a problem that has received little productive public attention in the past 40 years but has ballooned in size due to the COVID-19 pandemic. Much of the attention ME and Long Covid have received has tended to minimize the diseases and perpetuate misconceptions (Hsu, 2023; Yong, 2020a, 2020b).

Protesters continued to chant and hold signs while lying on the sidewalk in front of the White House. A group of secret service members and park police watched from across the street. After about 15–20 minutes of lying down on the hot cement, organizers decided to change tactics. This backup plan had been discussed in advance: protesters would move to one of the intersections nearby (since there is no vehicle traffic allowed on the street in front of the White House) and occupy a crosswalk, blocking traffic in the middle of bustling downtown Washington D.C. Protesters began to roll and walk toward the intersection. Those in the front of the group carried a large banner highlighting our cause. A few people helped pass along messages about pace between those in the front of our procession and those in the back so that we didn't leave anyone behind.

Unfortunately, as we moved toward the intersection, the downtown D.C. police realized what we were doing and diverted traffic so that there was nothing to block when we reached the intersection. Our group then decided to move to the next intersection. As we paused briefly to discuss strategy, someone encouraged a few faster people to go ahead of the others in the hopes of reaching the intersection before the police could divert traffic. Other experienced activists were adamant that we all stay together and leave no one behind. We followed the latter process. Because we moved slowly, the police again diverted traffic before we reached the intersection. As we reached the crosswalk, the subgroup engaging in civil disobedience spread out in the middle of the painted crosswalk area anyway. Standing, sitting, leaning, holding hands, linking arms, and holding signs, we blocked the intersection despite the lack of traffic. A few police officers stood politely across the street waiting for us to finish. It felt like participating in a waiting game that we knew we couldn't win; we wanted to be able to last long enough that they would arrest us, but lack of stamina is

FIGURE 9.2   Photograph (#MEAction, n.d.) of #MEAction protesters occupying the crosswalk in Washington D.C.

a core feature of our disability. Finally, we ended the protest and gathered
on a grassy lawn in the nearby park to rest and debrief (Figure 9.2).

Instead of the more compelling narrative and attention that might have
come from disrupting traffic in downtown D.C. or the visual image of protesters being arrested, spreading out in the empty intersection only attracted
attention from a few pedestrians who passed by and asked about our protest. While it may be unsurprising to those familiar with historical and
present-day U.S. racism and its intersection with policing that the police
were unwilling to arrest a group of visibly disabled, predominantly white
women, this situation is another potential example of the ways that activists can learn from participation: our experience illustrated that even when
bodies that are powerfully disabled by systems of capitalist productivity
and value, they can also wield a great deal of privilege. Experiences of disability are shaped by multiple, intersecting systems of oppression: disabled
people are more likely to be victims of violence than nondisabled people
and are overrepresented in arrests and prisons; that risk is compounded by
racialization which also increases these likelihoods along with the probability of harsher sentences (McCauley, 2017; Robin & McCoy, 2022). This
lesson is important because intersectionality is one of the core principles
of disability justice (see Sins Invalid, 2019) and is also used by scholars in
TPC to highlight the ways oppressions shape the field (Walton et al., 2019).

One of the things that I learned from this protest and the decision-making that occurred as we interacted with others' unpredictable responses is a lesson that I'm sure many disability-related movements have learned to consider: "success" is not solely defined by our ability to gain press attention or policy response, but also by our ability to foster community. In a disability context, an important part of that community-building work is creating "collective access," another principle of disability justice, by building a movement that includes people regardless of their level of function and leaves no one behind (Mingus, 2010; Sins Invalid, 2019). So, although we did not achieve our goal of blocking traffic in downtown DC, we did ensure that everyone who wanted to participate was included. This lesson is important because many social movements are often not designed with this kind of accessibility in mind. Many social movements have few or no participation options that are accessible to people with ME or Long Covid, particularly if their disease is moderate or severe.

This narrative of #MEAction's protest highlights a central problem facing activists and organizations focused on ME and Long Covid advocacy: some of the people most impacted by the lack of research and treatment for these diseases are also those for whom traditional activism, such as direct action, is inaccessible. The reason ME and Long Covid communities draw inspiration from HIV/AIDS activism is because the work of those activists and the subsequent developments in scientific understanding and treatment show that resources and policy can substantially improve life for people living with incurable[3] chronic disease (Brea & Staley, n.d.; Cooper, 2018; San Emeterio, 2022). However, one of the primary challenges that people with ME and Long Covid experience is that they have a dearth of energy *and* chronic symptoms that are exacerbated by exertion and/or sensory stimulation. As Emily Lim Rogers (2022) explains, this "recursive debility" presents a problem for activists, in which "debilitation blocks the means through which debilitation might end." That is, people with ME and Long Covid need to fight the invisibility of their suffering and the lack of awareness and attention to these diseases in research, but they have significant limits on their ability to engage in this advocacy and activism. However, designers, scholars, and students can learn a great deal about accessibility from people with ME and Long Covid and the disabled wisdom they use to create accessible activities for their community.

Overview of the Accessibility Problem/Challenge

While incapacitating fatigue is a central symptom of ME and Long Covid, the most pressing problem is more complex than fatigue. ME and Long Covid are both associated with a phenomenon called post-exertional malaise (PEM).[4] PEM involves "worsening of fatigue- and pain-related symptoms" after exertion (Appelman et al., 2024). While physical exertion such as exercise is certainly an important cause of PEM, activities that would not be considered "exertion" for people without ME or Long Covid can also cause PEM. Such activities include physical activities such as brushing teeth or showering; mental exertion such as concentrating, reading, or having a conversation; emotional exertion such as dealing with stress; and even experiencing strong stimuli such as bright light or loud noise. PEM is key to a diagnosis of ME and is experienced by a majority of people with Long Covid (Davis et al., 2023).

Every activity that someone with ME participates in not only takes from their already limited pool of resources for daily activity but also has the potential to cause worsening debilitation that could persist for months or even permanently alter the person's baseline. For example, after the 2022 protest in Washington DC, sociologist and *New York Times* columnist Zeynep Tufekci (2022) used brief daily phone and Zoom calls to follow up with protesters for two weeks as part of her work on a column about the protest. Her goal was to learn more about how participating in the protest affected people with ME and Long Covid afterward, since these after-effects are typically not perceptible to the public. She wrote that during the two weeks after the protest, she "watched many of [the protesters] deteriorate from the effects of postexertional malaise ... the same people who had managed to travel to the protest and had been so lucid and animated spoke to me from their beds and couches in halting, short sentences, frequently losing words and their train of thought. This is the largely invisible part of their disease." Because of these after-effects and the knowledge from patient communities that overexertion can lead to a decrease in baseline functionality, organizing protests comes with risks that the community must balance with the potential gains of participation. In addition to the risks of post-exertion symptom exacerbation for people with ME and Long Covid, traveling to the protest is also an expense that many people with these diseases can't afford because most people with ME and Long Covid are unable to work full time.

A common refrain among disability activists is "nothing about is without us," which emphasizes the importance of including disabled people in initiatives directed toward them (Bérubé, 2015; Costanza-Chock, 2020). This refrain responds to a long history of nondisabled people speaking for and/or claiming to represent disabled communities without including people from those communities in decision-making. As Sasha Costanza-Chock (2020) writes, "at its best, a design justice process is a form of community organizing ... wherever people face challenges, they are always already working to deal with those challenges" (p. 92). For ME and Long

Covid communities, inclusion and representation are complex because the severity of one's disease impacts whether they can participate in activism at all, what forms of participation are possible, and what risks come with participation. These challenges associated with functionality also intersect with axes of oppression that affect whether and how people with ME or Long Covid are connected to activist communities and campaigns, which I discuss in the next section using a specific organization, #MEAction, as a case study.

Design Case Study: #MEAction

#MEAction is an organization engaging in advocacy, public awareness-raising, and activism around myalgic encephalomyelitis (ME) and Long Covid. It was founded by Jennifer Brea and Beth Mazur in 2016. While #MEAction initially focused on ME, the organization shifted its focus in response to the COVID-19 pandemic in order to try to help people who were experiencing ongoing symptoms after acute COVID-19 infection ("Long Covid and ME," n.d.). The organization produced educational material and social media campaigns focused on educating people about how post-viral diseases work, including core symptoms such as post-exertional malaise ("Stop. Rest. Pace.," n.d.). #MEAction has since developed relationships with organizations focused on Long Covid such as Body Politic and Long Covid Justice ("#MEAction and Long Covid Justice," 2023). Some of the consultants, board members, and volunteers who work with #MEAction also have experience in HIV/AIDS activism and have brought that experience to their work with #MEAction (Ford, 2018; San Emeterio, 2022). In this section, I examine how #MEAction designs their #MillionsMissing campaigns in ways that consider community-specific circumstances and needs.

As a person with ME, I am a part of the community that some of these processes are designed for. I have also worked as a consultant with #MEAction and, as the earlier narrative indicates, I have participated in their campaigns. In 2022, I reached out to #MEAction to see if my experience studying social media and activism could be helpful for their work. They invited me to join their planning team for the protest in September 2022. My role involved thinking about how to introduce and facilitate the digital and social media component of the protest. I briefly discuss that work later in the chapter, but this section focuses more broadly on how different iterations of #MillionsMissing include and/or represent various groups of people with connections to ME and Long Covid.

Information on levels of severity and debilitation provides important context for the design work that #MEAction does. While all people with

ME and Long Covid experience disability that might affect their ability to participate in activism and advocacy, the level of function that people with ME and Long Covid experience varies, both person-to-person and from one day to the next. For example, people like me who would be considered "mild"[5] in relation to moderate or severe forms of ME are able to work full time but typically use up most of their energy on work and spend breaks like weekends and holidays recovering, which leaves little capacity for activism and advocacy. In addition to that general baseline functionality, symptoms like the aforementioned post-exertional malaise can mean that a person with mild ME or Long Covid who has overexerted herself might temporarily be unable to work or socialize while they experience symptom exacerbation (sometimes referred to as a "flare up" or "crash") that further debilitates them. Only about 13% of people with ME are able to work full-time. People with more moderate ME can't work full time due to their limited functionality and have a reduced ability to do many activities, including caring for themselves or concentrating. People with severe ME or Long Covid may not be able to leave their home or their bed, limiting their ability to do most things, including travel to in-person protests and possibly also participation on social media. Caregivers, too, spend extra time and effort supporting and caring for their loved ones with ME or Long Covid and therefore also have limited time and energy to invest in activism. Designing a protest for people with ME or Long Covid, then, involves acknowledging that some people will be completely excluded from participating due to the severity of their disease and most people in the community will have limited ability to participate regardless of the participation options.

Now, I turn to protest design to share examples of the ways that #ME-Action, the organization that created the #MillionsMissing protests, designs ways to represent people who can't be there and to allow people with varying levels of function to participate. The first #MillionsMissing protests involved creating displays of empty shoes in public spaces. For example, the Washington, D.C. #MilionsMissing protest, held on May 25, 2016, involved a display of empty shoes in front of the U.S. Department of Health and Human Services. The shoes were chosen to represent people who are "missing" from many aspects of their lives (e.g., from school, parenthood, work, hobbies, family life, and also from community and civic participation) due to myalgic encephalomyelitis. This tactic was taken up in different cities in the United States, Europe, Canada, Australia, and on social media. Additional #MillionsMissing protests occurred in September 2016, May 2017, May 2018, and May 2019, representing a growing number of locations. Not all of these protests followed the same format of placing shoes in front of a building associated with health policy and/or funding, but they did adopt the #MillionsMissing name and use the notion of "missingness" to draw attention to people who cannot participate in many activities. Another variation on this theme of "missing" people has involved creating protest signs that mimic the genres of "missing" posters and other campaigns (such as the information about missing children distributed on U.S. milk cartons in the 1980s).

In 2020, #MillionsMissing, like many other events and activities that spring, was shifted to a virtual format because of the COVID-19 pandemic. #MEAction produced a short film titled "I got a virus, I didn't die, but I never recovered. We are the #MillionsMissing," released on May 10, 2020. The short film draws attention to the long-term consequences that viruses can have. Many cases of ME are triggered by viruses, and previous viral outbreaks have led to outbreaks of ME. Therefore, people with ME and experts on ME were some of the first to raise awareness about the possibility that the pandemic would lead to an upsurge in chronic disease.

In 2021 and 2022, #MEAction hosted two #MillionsMissing global virtual events using the hashtags #MillionsMore and #YouAreNotAlone to reach out to people experiencing Long Covid. Both virtual events were recorded webinars featuring a range of people with ME and Long Covid, as well as prominent experts, disability representatives, activists, and journalists. The events were recorded and posted to YouTube, increasing the accessibility of the events for many people, such as those who were unable to concentrate on the entire live event (Figure 9.3).

In September 2022, #MillionsMissing returned to in-person protests with the demonstration and civil disobedience described earlier in this chapter. In addition to the demonstration in Washington D.C., #MEAction created a virtual toolkit designed for people of varying energy levels to participate in whatever way they were able to. This was part of the work that I consulted with #MEAction on. We created a toolkit describing different participation options, using energy levels to organize the participation options. The actions were divided into five sections, each labeled with one to five spoons (a reference to blogger Christine Miserandino's "spoon theory," a popular story about what it's like to live with limited energy and to have to plan one's day and activities around those limitations). Within each section were options for people with and without social media to participate. For example, the one-spoon action for people with social media involved sharing, retweeting, and/or commenting on material shared by #MEAction. The four-spoon action for people without social media was to email a letter to President Biden that incorporates the participant's own story and/or photo. The virtual materials created and shared by #MEAction also included training on storytelling and taking photos, which were released well in advance of the protest date, as well as tips for increasing

#MEAction Network
@MEActNet
•••

#MillionsMissing is Monday! For those taking action from home, our Activism From Home Toolkit is filled w/ actions to take depending on your energy levels using the spoon theory. You can choose to do 1 action or you can do them all! See all actions below!
ow.ly/ECZu50KIIQC

to

#MillionsMissing Activism from Home

Social Media Actions organized by a scale of one to five spoons. Join us on September 19th.

ALT

2022 #MILLIONS MISSING

FIGURE 9.3 Screenshot of #MEAction tweet introducing the #MillionsMissing 2022 Activism from Home Toolkit.

social media connections to people with ME and Long Covid in order to encourage circulation of the material among the community. The toolkit and other materials emphasized that all of the actions, whether in-person or virtual, were important. As Holly Latham, #MEAction's social media manager wrote on Twitter (now known as X), "Remember that if you can do nothing on [the protest day], you are just as valuable & needed a part of our community as any other community member! If you retweet 1 tweet or put your body on the line to be arrested, we are grateful for you & are all an integral part of this movement" (@MEActNet, 2022). Through these materials, organizers, and activists created roles for community members of different severity levels and accounted for the fluctuating nature of chronic diseases, which some activists refer to as "dynamic disability" (see Abayomi-Paul, 2022; Benness, 2019).

#MillionsMissing 2023 took a new format. #MEAction created a display of 500 cots on the National Mall in Washington D.C. The display built on the previous tactic of using physical objects to represent people who cannot be present, but this time with cots covered in red blankets. In the weeks leading up to the event, people with ME and Long Covid decorated and mailed pillowcases to #MEAction for display on the cots. The pillowcases provided people with a way to contribute their own message to the protest without traveling to attend in person. #MEAction also hosted a press conference in front of the display. One of the goals of the press conference was to highlight the perspectives and experiences of people and experts whose communities have been underrepresented in ME and Long Covid activism and advocacy. The speaker lineup included Dr. Margot Gage, Mary Dimmock, Dr. V. Jo Hsu, Jaime Seltzer, Gabriel San Emeterio, Morgan Stephens, Dr. Dona Kim Murphey, and Brooke Keaton.

#MillionsMissing 2024's theme was "Teach Me, Treat Me." This theme involved outreach to local hospitals and healthcare providers to educate medical professionals about ME and Long Covid and build community partnerships. While many previous #MillionsMissing campaigns focused on actions designed to attract the attention of relevant governmental agencies or policymakers (such as the protest in front of the U.S. Department of Health and Human Services), this protest took a different approach by encouraging people with ME and Long Covid to develop events at local hospitals and health providers that publicized a continuing medical education course on ME and Long Covid. This design responds directly to another problem that ME and Long Covid communities experience: the lack of clinician knowledge about the pathophysiology of ME and Long Covid and corresponding ways to manage symptoms. Prior to the COVID-19 pandemic, very few medical schools (about 6%, according to #MEAction) taught students about ME. Because symptom management is complex, medical systems and practitioners were not prepared for the huge surge in the number of people with ME and Long Covid. Since typical advice from doctors, such as advice about exercise, can be harmful to people with ME and Long Covid if not approached carefully, practitioner education can increase the number of people receiving appropriate diagnoses and/or medical advice and reduce harm to people with these diseases.

The bodies of people with Long Covid and ME become sites of critique and resistance, not by choice but because their bodies literally cannot do the things that are aligned with human values under neoliberal capitalism. As their bodies refuse these demands, people with these diseases often face problems ranging from lack of support to gaslighting and dismissal. However, these communities have shown incredible creativity and perseverance in their activism and advocacy. These campaigns are by no means perfect (which is true of all activists and people), but they offer insight into the ways that community-engaged design can make creative use of technology and multimodal rhetoric to address the needs of marginalized communities and develop activities that are accessible to people with different experiences of debility.

Design and Reflection Activities for the Classroom

Pre-activity Reflection Questions

The purpose of this reflection is to cultivate some connection to people with ME and Long Covid through your own experience of illness or injury. However, please keep in mind that experiencing temporary illness or injury is not the same as experiencing incurable chronic disease.

1 When was the last time you experienced illness or injury that affected your ability to carry out your usual activities?
2 How did you feel? What symptoms did you experience?
3 How did these feelings and symptoms interfere with your ability to do things? How did you have to adjust your normal activities in order to accommodate your illness? Would it be possible to make this adjustment permanently? Why or why not?

Learning Activities

1 Choose one or more of the following resources to learn more about ME and/or Long Covid:

   a ME-pedia's entry on myalgic encephalomyelitis, including the section titled "Signs and Symptoms."
   b One of #MEAction's Pacing and Management Guides for people with ME and Long Covid. Pay special attention to the section titled "Exertion ≠ Exercise"
   c "How Managers Can Support Employees with Long Covid" by Fiona Lowenstein
   d *Unrest*, a documentary about ME that is available for free on YouTube.

2 Review the website storyofmillionsmissing.org for photos and descriptions of #MillionsMissing protests.

  a Review several pillowcases sent in by people with ME and Long Covid. What do you notice about these messages and designs?
  b Use the "press" tab to view at least one speech from the press conference. (Or, review the live broadcast of the press conference, available on Twitter/X.)

   • What did you learn from this speaker?
   • Who are they speaking to?
   • What do they say?
   • Why is their perspective important?

After Learning More, Reflect on the Following Questions

1 What did you learn about ME and Long Covid? How do people with ME and Long Covid experience the world differently than people who do not have ME or Long Covid?
2 How were your recent experiences of illness similar to and different from the symptoms described in the materials on ME and Long Covid?

  a What were you able and unable to do? What did you need from others at work, school, or home?
  b Or, if you are someone with ME and/or Long Covid, were you familiar with this information? How do you (or can you) use this information in your daily life?

Designing for People with ME and Long Covid

Imagine that a person with ME or Long Covid wants to participate in an advocacy campaign, community event, or another activity such as this class. Use the questions below to consider whether their participation would be possible, what changes would make the participation more accessible to them, and what effects their participation might have.

1 What participation options are available? How do people learn about this activity? How do they participate? How long does it last? What are the environmental conditions? Will there be breaks? Will there be space and/or support for mobility aids?
2 What kinds of *exertion* would the activity involve? Can any of this exertion be minimized or eliminated? How?
3 How can the activity organizers create community and social support for people with ME and Long Covid (or other dynamic disabilities)

who want to participate? What kinds of planning and resources would be needed for this work?

4 What resources would be required to make these options available? Think about things like time, expertise, material resources (like space or equipment), and social resourcess (e.g., community connections).

Notes

1 *Activist Literacies* p. 156.
2 For examples of these connections, see Cooper (2018) and Brea and Staley (n.d.).
3 Or once thought incurable, since there are several people whose HIV has been cured after receiving stem cell transplants as part of cancer treatment.
4 This phenomenon is also (less commonly) called "post-exertional symptom exacerbation" or "post-exertional neuroimmune exhaustion."
5 Quotation marks here indicate a relative term: "mild" does not mean that one's overall experience of disease and disability is mild, but rather that their experience of ME or Long Covid is mild in comparison to other people with ME and Long Covid.

References

#MEAction [@MEActNet]. (2022, September 17). *#MillionsMissing is Monday!* Twitter. https://twitter.com/MEActNet/status/1571228672685756417

#MEAction and Long COVID Justice Issue Press Statement About NIH Failure of Long Covid. (2023, August 23). https://www.meaction.net/2023/08/23/meaction-issues-press-statement-about-nih-failure-of-long-covid/

#MEAction. (n.d.) *Long Covid and ME: Understanding the Connection.* #MEAction. https://www.meaction.net/long-covid-me-understanding-the-connection/

#MEAction. (n.d.) [Photo of #MillionsMissing protestors in front of White House]. SmugMug. Retrieved May 31, 2024. https://meaction.smugmug.com/MillionsMissing-2022-1/MillionsMissing-2022/i-kDXtRKL/A

#MEAction. (n.d.) [Photo of #MillionsMissing protestors occupying crosswalk]. SmugMug. Retrieved May 31, 2024. https://meaction.smugmug.com/MillionsMissing-2022-1/MillionsMissing-2022/i-kDXtRKL/A

Abayomi-Paul, T. [@Tinu]. (2022, September 18). *This is so inclusive and accessible. A sound model for maximum participation in activism by energy level* [Quote Tweet; link to #MEAction tweet with image and link to Activism from Home Toolkit.]. Twitter. https://twitter.com/Tinu/status/1571557329161658371

Appelman, B., Charlton, B. T., Goulding, R. P., Kerkhoff, T. J., Breedveld, E. A., Noort, W., Offringa, C., Bloemers, F. W., van Weeghel, M., Schomakers, B. V., Coehlo, P., Posthuma, J. J., Aronica, E., Wiersinga, W. J., van Vugt, M., & Wüst, R. C. I. (2024). Muscle abnormalities worsen after post-exertional malaise in long Covid. *Nature Communications, 15*, 1–15, https://doi.org/10.1038/s41467-023-44432-3

Benness, B. (2019, December 8). *My disability is dynamic.* Medium. https://medium.com/age-of-awareness/my-disability-is-dynamic-bc2a619fcc1

Bérubé, M. (2015). Representation. In R. Adams, B. Reiss, & D. Serlin (Eds.), *Keywords for disability studies* (pp. 151–154). NYU Press.

Brea, J. (Director). (2017). *Unrest*. [Film]. Shella Films.

Brea, J., & Staley, P. (n.d.). Lessons from the AIDS movement: Peter Staley interview transcripts. *#MEAction*. https://www.meaction.net/how-to-survive-a-plague/peter-staley-transcript/

Callard, F., & Perego, E. (2021). How and why patients made long Covid. *Social Science & Medicine, 268*. https://doi.org/10.1016/j.socscimed.2020.113426

Cooper, G. (2018, February 19). What ME activists can learn from the AIDS crisis. *#MEAction*. https://www.meaction.net/2018/02/19/what-me-activists-can-learn-from-the-aids-crisis/

Costanza-Chock, S. (2020). *Design justice: Community-led practices to build the worlds we need*. MIT Press.

Davis, H. E., McCorkell, L., Vogel, J. M., & Topol, E. J. (2023). Long COVID: Major findings, mechanisms and recommendations, *Nature Reviews Microbiology, 21*, 133–146, https://doi.org/10.1038/s41579-022-00846-2

Devlin, H. (2023, August 23). Women with ME tend to have more symptoms than men, study suggests. *The Guardian*. https://www.theguardian.com/society/2023/aug/24/women-with-me-tend-to-have-more-symptoms-than-men-study-suggests

Dolmage, J. T. (2014). *Disability rhetoric*. Syracuse University Press.

Ford, O. G. (2018, May 9). Using lessons from HIV activism, people with 'chronic fatigue syndrome' fight for the #MillionsMissing. *The Body*. https://www.thebody.com/article/people-with-chronic-fatigue-millionsmissing

Fricker, M. (2007). *Epistemic injustice: Power and the ethics of knowing*. Oxford University Press.

Hamraie, A. (2017). *Building access: Universal design and the politics of disability*. University of Minnesota Press.

Hsu, V. J. (2023). Framing the activists: Gender, race, and rhetorical disability in contested illnesses. *Quarterly Journal of Speech, 110*(2), 198–220. https://doi.org/10.1080/00335630.2023.2291895

Mancini, D. M., Brunjes, D. L., Lala, A., Trivieri, M. G., Contreras, J. P., & Natelson, B. H. (2021). Use of cardiopulmonary stress testing for patients with unexplained dyspnea post-coronavirus disease. *JACC. Heart Failure, 9*(12), 927–937.

McCauley, E. J. (2017). The cumulative probability of arrest by age 28 years in the United States by disability status, race/ethnicity, and gender. *American Journal of Public Health, 107*(12), 1977–1981. https://doi.org/10.2105/AJPH.2017.304095

Mingus, M. (2010, August 23). Reflections on an opening: Disability justice and creating collective access in Detroit. *Leaving Evidence*. https://leavingevidence.wordpress.com/2010/08/23/reflections-on-an-opening-disability-justice-and-creating-collective-access-in-detroit/

Nish, J. (2022). *Activist literacies: Transnational feminisms and social media rhetorics*. University of South Carolina Press.

O'Rourke, M. (2023). *The invisible kingdom: Reimagining chronic illness*. Penguin Random House.

Phillips, S., & Williams, M. A. (2023, September 14). Long Covid is a new name for an old syndrome. *STAT*. https://www.statnews.com/2023/09/14/long-covid-me-cfs-myalgic-encephalomyelitis-chronic-fatigue/

Robin, L., & McCoy, E. F. (2022, August 29). The criminal legal system fails to address Black disabled people's intersectional identities. *Urban Wire*. https://www.urban.org/urban-wire/criminal-legal-system-fails-address-black-disabled-peoples-intersectional-identities

Rogers, E. L. (2022). Recursive debility: Symptoms, patient activism, and the incomplete medicalization of ME/CFS. *Medical Anthropology Quarterly*, *36*(3), 412–428. https://doi.org/10.1111/maq.12701

San Emeterio, G. (2022, December 8). I live at the intersection of pandemics: HIV, ME/CFS & Long Covid. #MEAction. https://www.meaction. net/2022/12/08/i-live-at-the-intersection-of-pandemics/

Sins Invalid (2019). *Skin, tooth, and bone: The basis of movement is our people* (2nd ed.). Author.

Stop. Rest. Pace. (n.d.). *#MEAction*. https://www.meaction.net/stoprestpace/

Tham, J. C. K. (2021). *Design thinking in technical communication: Solving problems through making and collaboration*. Routledge.

Tillman, A. (2022, September 20). #MillionsMissing press hits are rolling in! *#MEAction*. https://www.meaction.net/2022/09/20/millionsmissing-press-hits-rolling-in/

Tufekci, Z. (2022, October 27). Protesters so ill, they couldn't get arrested. *The New York Times*. https://www.nytimes.com/2022/10/27/opinion/me-cfs-long-covid.html

Walton, R., Moore, K. R., & Jones, N. N. (2019). *Technical communication after the social justice turn*. Routledge.

Yong, E. (2020a, September 9). America is trapped in a pandemic spiral. *The Atlantic*. https://www.theatlantic.com/health/archive/2020/09/pandemic-intuition-nightmare-spiral-winter/616204/

Yong, E. (2020b, June 4). Covid-19 can last for several months. *The Atlantic*. https://www.theatlantic.com/health/archive/2020/06/covid-19-coronavirus-longterm-symptoms-months/612679/

10

RESISTING THE DATAFICATION OF INJUSTICE THROUGH COLLABORATIVE DESIGN AND TRANSLATION IN A LOCAL MUSEUM EXHIBIT

Laura Gonzales, Valentina Sierra-Niño, and Robin Lewy

Introduction: What Is Language Access?

Language access is a Civil Right protected by several legal documents, including the Civil Rights Act of 1964 and Executive Order 13166. These laws dictate that all entities receiving federal funding must provide avenues for multilingual community members who speak languages other than English to meaningfully participate in organizational events. As part of these guidelines, organizations receiving federal funding are required to develop a language access plan, which is a document that clearly outlines the organization's commitments and protocols for providing avenues of meaningful participation for multilingual community members. These technical documents, including the laws and the language access policies, are intended to protect the rights of multilingual individuals, providing them with access to services in their community. However, as technical communication scholars such as Jones and Williams (2018) demonstrate, technical documents, even those intended to protect individuals, are not always created with benevolent intent. Indeed, technical documents are often used to further oppress the very communities they claim to serve.

In our own community, a semi-rural town in North Central Florida, we have noticed that language access policies, protocols, and documents, are sometimes used as a facade to feign accessibility and inclusion (Welcoming America, 2024). For example, local immigrant activist groups have repeatedly stated that our multilingual community members are still not provided with access to language access services in our local schools. Multilingual parents in our community have filed several complaints with the

DOI: 10.4324/9781003469995-13

local school board, stating that they do not have a mechanism for communicating with their children's school to learn about changing policies, bus routes, and other school-related opportunities. Technical documents, like language access policies, are used to feign compliance with language access laws, while parents and other multilingual community members remain under-served and under-represented in their local community.

Redressing the Datafication of Injustice through Collaborative Design

Since 2018, following a tragic domestic violence incident involving an immigrant family, local organizations in our community have been advocating for language access policies to be implemented throughout our local city and county (Gainesville Immigrant Neighbor Inclusion Initiative, 2024). In 2022, a local group of immigrant-focused activists, known as the Gainesville Immigrant Neighbor Inclusion Initiative (GINI), developed several strategies to foster language access efforts across the community. The group conducted a survey to better understand the language diversity embedded in the community, and they developed a document, the "GINI Immigrant Inclusion Blueprint," which outlines the languages most commonly spoken by survey respondents in our region and provides guidelines for how organizations can work with these communities to increase the participation opportunities for multilingual immigrants throughout the city and county.

While this Blueprint and the group's efforts have made important progress in enhancing immigrant inclusion in our community, grievances and complaints from multilingual families continue to demonstrate the need for further action. Institutions with the power to provide resources and funding for language access continue claiming that we need to provide more data, and more evidence to demonstrate the need for language access resources in a town often presumed to be linguistically homogeneous. Despite the fact that our community is situated in Florida, a city where immigrants accounted for 24 percent of the population growth between 2014 and 2019 (New American Economy, 2021), local immigrant activist groups continue receiving pushback from county and city officials who claim there is not enough data to support the need for additional language access resources.

In *Viral Justice: How we Grow the World we Want*, Benjamin (2022) names this consistent request for data an intentional strategy to perpetuate injustice and prevent change. Using the example of the long-documented health detriments imposed upon Black women through misogynoir, Benjamin (2022) defines the datafication of injustice as "the hunt for more and more data about things we already know much about" (p. 35). Instead of consistently requesting more and more data to further "prove" that systemic injustices exist, Benjamin (2022) argues that "the research community needs to reckon with how our work contributes to *structural gaslighting*" (p. 35). As Benjamin (2022) argues, we don't need one more study to "prove" that Black women experience interlocking structural oppressions that deter their lifespan. Drawing on this work, we as authors of this article and members of our local immigrant activist coalition, also recognize, in a different capacity, that our community did not and does not need another survey to "prove" that multilingual people exist in our context, and that these community members experience harm in our community because they do not speak English. Indeed, we are close to the multilingual members of our community, with one author of this paper having served immigrants in the region for over 30 years. We know the stories of injustice that our multilingual immigrants share, and we also recognize that it's not easy for them to share these stories repeatedly with local officials. Furthermore, we recognize that the Census and other existing data do not adequately represent all multilingual immigrants in any given context, as vulnerable multilingual populations have reasons not to trust established surveys and documentation.

Thus, in 2021, we came together with our community allies to attempt another strategy: highlighting multilingualism through collaborative design. For under-resourced, yet resilient, communities, positive and negative emotions may intertwine in situations of stress (Ong et al., 2006). Emotions, as noted by Salomon (1989), are part of our "first line of recognition of what's right and wrong, justice and injustice" (p. 346). We set out to highlight multilingualism by delving into the joys and challenges associated with languages in our community. We consider that tension between the negative and positive experiences and emotions is critical to portray injustice in a way that inspires change but also embraces and honors the cultural and social values people assign to their languages.

With the support of our local history museum, we designed a museum exhibit that illustrated, through imagery and words in multiple languages, what language means to the multilingual immigrant members of our community. We interviewed 80 community members, asking them to tell us about their language, to share their immigration story, and to also share with us examples of some experiences that they've had when speaking their own languages in our community. We then transformed this interview data into visual interactive collages that highlighted the stories our participants told during their interviews. We used birds and plants as metaphors for migration, including birds and plants from our participants' home communities in each collage (Sierra Niño, 2024). The goal of this exhibit was to push back on the datafication of injustice by using design to visually illustrate the experience of being multilingual in our local context in a way that embraced the complexities of people's experiences.

Using visuals and other modalities to expand access and share complex narratives is a common practice in both technical communication and writing studies. For example, in her work on multimodal captioning, Butler (2016) explains, "the synchronicity of modes—with sound accompanying visual, gestural, and spatial movement—reinforces the value of designing access to various modes in multimodal compositions" (n.p.). While Butler (2016) focuses on the affordances of embodied movement as a form of multimodal expression, many other scholars highlight the value of layering various modalities to expand engagement, particularly in situations where audiences speak multiple languages. Gonzales and Butler (2020) argue that "communicative practices are fluid, emergent, always in flux, and that, therefore, students should be encouraged to write through various modalities that reflect the flexibility of contemporary communicative practices" (n.p.). Expanding definitions of multimodality to include digital interfaces, Haywood (2022) explains, "from blogging to the use of #BlackLivesMatter hashtags, digital and Internet spaces grant Black women the ability to exist, write, and work in ways that significantly add to the varied and extensive writing and rhetorical histories that they carry" (p. 20). In other words, in communities where standardized, White American English is not the only form of communication, multimodal forms of expression continue to provide avenues for increasing participation, engagement, and access. Thus, using visual collages, interview quotes in visual form, and other data points to design a collaborative exhibit, provided us with the opportunity to reflect the stories of our community in an engaging manner that brought together multiple stakeholders. In this chapter, we discuss how designing this multilingual museum exhibit helped foster and shape conversations about the sense of justice and language access in our community.

Launching the Exhibit: We Are Here

On May 19th, 2023, we brought together over 120 community members to our local history museum for the launch of our art exhibit, titled, *We are Here: Stories from Multilingual Speakers in North Central Florida*. The exhibit included several visual representations of the stories shared by our participants during their interviews. In addition to the collages, for example, the main entryway of the exhibit included the exhibit title alongside quotes from the interview that describe how our multilingual community members feel about language access in their local context. Figure 10.1 is an image of the exhibit title with the surrounding quotes.

FIGURE 10.1 Exhibit title with quotes.

As evidenced in Figure 10.1, when describing their experiences speaking their language in the community, our participants made statements that echoed the challenges of being multilingual, including, "losing my power," "we feel violated," "not even seen," "feel vulnerable," and "threatened." Participants also shared positive things related to speaking their languages, including, "the language that I love" and "my nurturing language." The goal of the exhibit was to demonstrate that linguistic experiences are dynamic. People can love speaking their home languages in the community, and they can also feel vulnerable and challenged when they cannot communicate comfortably in English. Stories of migration are often filled with both joy and pain. In our exhibit, we wanted to portray all parts of the immigration journey, giving the opportunity for our local community members to express all parts of their identities and their experiences.

Increasing Representation

The opening reception for the exhibit included speeches by members of the community who were also interviewed and included in the exhibit itself. Community members launched the exhibit by sharing their experiences both participating in this project and navigating language access in their city and county. Speakers at the launch event came from multiple different countries, including Kenya, Colombia, Bolivia, Senegal, France, Côted'Ivoire (Ivory Coast), Haiti, Jamaica, Guatemala, and the US. Within these home countries, these guests spoke multiple different languages. For example, a group of community members from Guatemala spoke both Spanish and Q'anjob'al, an Indigenous language from Guatemala. The group of speakers included parents and several children, all of whom explained that they speak their Indigenous language at home to preserve their culture and heritage. Furthermore, the children in this family were proud to share that they are trilingual, speaking Q'anjob'al, Spanish, and now, English, following their family's immigration to the United States.

Too often, conversations about language access centralize white immigrants who speak white, European descendant languages. Yet, as we learned through our interviews with community members, multilingual speakers of color, and in particular, Black and Indigenous immigrants in Florida, face intersecting injustices when speaking their heritage languages. During one of our interviews, a participant who speaks Haitian Creole, French, and English shared her experiences navigating healthcare as a Black multilingual immigrant woman, stating, "There is a racist belied circulating in the US that the Black population is only composed of African Americans, This has, in turn, created an assumption that Black people only speak English, so they don't need access to language services." Another participant, who speaks Bengali, Hindi, and Urdu and was learning Spanish, shared that "Languages like Spanish or Italian, or French, or German are kind of looked up to. But when I say I speak four languages, but three of them are not European or Western, I don't feel like I'm valued." A primary goal of our exhibit was to showcase the multiple identities embedded within the broad labels, "multilingual," and "immigrant," and to demonstrate that conversations about language should always include conversations about race, privilege, and power.

Visualizing Data to Shape Stories

The primary elements of the exhibit were 10 interactive collages that included quotes from a community member interview as well as images of birds and flowers from that participant's home country. The content of the collage was written in the participant's heritage language. Exhibit visitors could access a translation of that content by pulling a tab on the collage. For example, Figure 10.2 is a collage created for a participant from Guatemala who spoke Q'anjob'al. The collage includes images of the Quetzal, the national bird of Guatemala, as well as a story from that participant in Q'anjob'al. As evidenced in Figure 10.2, each collage had a tab that stated, "For English, pull Here," as well as a tab that stated, "For the meaning of the images, pull here." These tabs allowed us to centralize the multilingual content in the exhibit, while also providing English-speaking attendees of the exhibit with the opportunity to experience what it can feel like to engage with a text in a language you cannot understand. By having to

FIGURE 10.2 Q'anjob'al collage with tabs for translation.

pull on the tabs to access the exhibit content, English-speaking audiences who could not understand the primary language of the collage could get a glimpse into the additional labor that multilingual community members have to engage in to translate information into their preferred languages. This form of embodied multimodality, encouraging attendees to engage with the exhibit with their bodies by pulling on tabs to learn more, further echoes Butler's (2016) emphasis on synchronicity and movement as an important aspect of composition.

In addition to the visual collages, we used design to (re)tell the story of language history in North Central Florida. The exhibit included a visual timeline that told the history of the languages spoken in Gainesville and in Alachua County. As we note on this timeline, while we might assume that language access needs and efforts begin with immigration, the truth is that English is not a native language to Florida (or to the US). The first language spoken in Gainesville was Timucua, spoken by the Timucua people, the principal Native American population of North Central Florida. In 1513, Timucua's land (present-day Florida) was invaded by Spain under the colonial leadership of Juan Ponce de León. At this time, the Spanish language was used to eradicate the Timacuan languages. Thus, Spanish and Timucua were spoken in Florida before English. Furthermore, in 1526, Spanish colonizers trapped and enslaved people from many African countries and forcefully brought them to "La Florida" to continue building the Spanish empire. Enslaved people, brought to Florida against their will, spoke multiple African languages and language variants, which then also made contact with Spanish, Timucuan languages, and later, English. This linguistic connection, through an act of forced displacement and enslavement, was part of the foundation of what later became named African American Language (AAL), which is prominently spoken in Florida and across the US, and continues shifting through ongoing physical and linguistic migrations. Using a technical document, such as a timeline, allowed us to illustrate the fact that English was not spoken in Florida until the end of the Seven Years' War in 1763, when the British Empire took control over Florida.

After exhibit attendees view the timeline of Florida's languages, they are guided to walk by a section of the exhibit that includes letters, memos, and other documents with most of the written content blacked out using a permanent marker. These documents included letters, bills, and notices that we collected from our multilingual community members. These letters were sent to our community members' homes in English and without translation, leaving community members with the task of finding a translation for this information. In the exhibit, we used the permanent marker to cover the content, simulating what it might feel like for people to receive these notices without being able to understand what is in them. For example, we included letters with large numbers and dollar signs that were sent to community members in an effort to collect funds through medical billing. Community members who do not speak English were not able to understand why they were being charged, and they could not understand information about potential payment plans and other resources that they may have been entitled to. In this section of the exhibit, we included the following statement: "Multilingual community members who are not provided with language access services are not able to exercise their rights in many aspects of our community. Language access impacts multiple sectors of community involvement, including health, education, and safety. Imagine receiving notices in the mail that seem urgent, but which you cannot understand. What would you do? Who would you ask for help? How would you feel? Reflect on these questions as you attempt to read these documents."

Through this aspect of the exhibit, we demonstrated what Jones and Williams (2018) refer to as the "technologies of disenfranchisement" that dominate multiple technical documents in American society. Literacy tests, medical bills, memos, and even event invitations, when provided in languages that are inaccessible to their target audience, can perpetuate oppression. At the same time, multimodal design practices, such as the collages, the timeline, and the embodied experience that we curated for individuals navigating the exhibit, helped to both highlight and perhaps minimally address this disenfranchisement. Through this exhibit, we sought to make visible, in literal terms, what multilingual communicators may experience when receiving technical communication documents that they cannot understand, but which also seem to demand an urgent response.

Furthering the Conversation

While this exhibit was in part designed to resist the push for more data about the need for language access, we also wanted to incorporate opportunities for attendees to contribute their own stories to the exhibit. In the back of the exhibit room, we included an interactive section, where participants could share where they are from and what they experience when speaking their languages in our community. In this section, we included a globe alongside pins, where attendees were asked to add a pin to the globe on top of the country/region that they are from. Furthermore, we provided a set of postcards in this same area, where we asked participants to write down where they are from, what languages they speak, and why their languages are meaningful to them. We also asked participants to share both experiences of joy and challenges they have faced when speaking their heritage languages in our community. Figure 10.3 is an image of the globe and two of the postcards completed by our exhibit attendees.

FIGURE 10.3   Globe and postcards at the interactive section of the exhibit.

On the final day of the exhibit's viewing, the globe contained 79 pins in 79 different countries and regions, thus making visible the vast diversity of heritage countries and languages that interacted with a small local history museum in a semi-rural town in Florida. The postcards contained dozens of stories from community members who shared their experiences speaking multiple different languages in the community. For example, one postcard, written by an attendee who speaks Spanish and English, reads, "No solo me han discriminado por el idioma si no mi físico latino. Aunque nací en California y me crié en Costa Rica. Siempre hay prejuicios" ("I've been discriminated against not only because of my language but also because of my Latino appearance. Even though I was born in California and raised in Costa Rica. There are always prejudices"). Another postcard, written by a participant who speaks English and Arabic, reads, "My languages are meaningful to me because they connect me with my culture and allow me to be tied to my roots." A speaker of Russian, Spanish, and English wrote that language "is a way to bridge dives and help make connections for those who cannot yet help themselves."

In their responses, some visitors expressed pride in their languages. However, they also shared how in situations of limited language access and lack of respect towards language diversity, this source of pride transforms into a vulnerability trigger. For example, a participant who speaks Gullah mentioned, "I have been told by some that my language doesn't exist, and some have said that I or my family members sound ignorant when we speak. I will not be ashamed of my culture and language." Another participant who speaks Bangla wrote, "my language is my greatest insecurity but also my connection to my culture, I love Florida but I've never felt so isolated in a place that should be my home." These two examples show feelings of discrimination, loneliness, and even dehumanization when others deny participants' culture.

Some participants shared how people judging them for speaking languages other than English created fear and affected the ways they felt and behaved when speaking their languages. One of the postcards read, "I internalized that I can't speak my native tongue, even if I'm not discriminated against, I try not to speak my language outside due to fear of standing out, being different". Similarly, another participant wrote, "Sometimes it is a little scary speaking Chinese with my parents on the phone in public, people will watch me or stare."

Participants also used the postcards to reflect on their own identities as English speakers. For example, one attendee described the languages she speaks as "English with AAL Flava." In her answer to the question, "What challenges or discrimination have you experienced when speaking your language in Florida," the attendee wrote, "As a Black woman English speaker my language and body/identity cannot be separated. When I speak, my whole culture flows through my words. Each pronunciation, beat, accent, etc. is up for interpretation. Is it too Black, too White, or not enough?" In sharing her story, this attendee, and therefore, exhibit contributor, showcases the importance of considering language, identity, gender, and race, together, rather than in isolation (see also, Browdy & Milu, 2022).

Pedagogical Implications

While the museum exhibit we describe in this chapter is situated in a specific community, the design activities we incorporated in developing the exhibit's experience can have broad applications for technical communication, design, and community engagement practitioners. For example, technical communication researchers should consider the role that language plays in all of our community projects. Investigating the linguistic histories and practices of our communities, and intentionally developing avenues for community members to discuss these experiences, can provide important perspectives about the role that identity, culture, race, and positionality play in all research projects. Furthermore, developing avenues to incorporate multimodal elements, including visuals, embodied movement, and digital technologies, into community projects can help bridge communication gaps among researchers and community members. This bridge in communication is particularly important when working with community members to resist the "datafication of injustice" (Benjamin, 2022) by leveraging ways to share difficult stories in accessible formats. As this project demonstrates, language can provide an avenue for fostering engagement and collaboration, particularly when working with communities who have always worked outside the boundaries of standardized White languages.

Conclusion

The exhibit we describe in this chapter was on display for 9 months, from May 2023 to February 2024. During this time, hundreds of visitors engaged with the exhibit's material, reading the technical documentation, interacting with the multilingual collages, and contributing to the exhibit through the learning exercises provided through the inclusion of the globe with pins and the postcards with reflective questions. Furthermore, as designers of the exhibit, we were able to host events at this museum space, including the exhibit launch, a collage-making workshop where we invited community members to create their own visual stories about language, and a storytelling event where we invited local multilingual immigrant community members to share their stories of immigration and brainstorm strategies for making out community more inclusive to multilingual immigrants.

While the exhibit has come to a close, the work of fostering language access and immigrant inclusion continues. Currently, our state continues implementing xenophobic and racist policies that limit what organizations can do to support the most vulnerable members of our immigrant community, including people without status, multilingual immigrants of color, and multilingual immigrants with disabilities. Through these continuous oppressions, advocates for immigrant inclusion continue innovating ways of strategically supporting our multilingual immigrant neighbors. Bringing together translation, technical communication, and design continues to be an integral and important way to both expand the impact of immigrant stories and experiences and resist the incessant datafication of injustice. Through community, we continue developing mechanisms for showcasing what we already know: that immigrants who do not speak English have the rights to access information in their community.

References

Benjamin, R. (2022). *Viral justice: How we grow the world we want.* Princeton University Press.

Browdy, R., & Milu, E. (2022). Global Black rhetorics: A new framework for engaging African and Afro-Diasporic rhetorical traditions. *Rhetoric Society Quarterly, 52*(3), 219–241.

Butler, J. (2016). Where access meets multimodality: The case of ASL music videos. *Kairos: A Journal of Rhetoric, Technology, and Pedagogy, 21*(1). https://kairos.technorhetoric.net/21.1/topoi/butler/index.html

Gainesville Immigrant Neighbor Inclusion Initiative. (2024, February 11). *Gainesville Immigrant Neighbor Inclusion Initiative.* https://gini-initiative.org/index.html

Gonzales, L., & Butler, J. (2020, June). Working toward social justice through multilingualism, multimodality, and accessibility in writing classrooms. *Composition Forum, 44.* https://compositionforum.com/issue/44/multilingualism.php

Haywood, C. (2022). Developing a Black feminist research ethic: A methodological approach to research in digital spaces. In V. Del Hierro & C. VanKooten (Eds.), *Methods and methodologies for research in digital writing and rhetoric: Centering positionality in computers and writing scholarship.* Vol. 2. (pp. 29–44). The WAC Clearinghouse; University Press of Colorado.

Jones, N. N., & Williams, M. F. (2018). Technologies of disenfranchisement: Literacy tests and black voters in the US from 1890 to 1965. *Technical Communication, 65*(4), 371–386.

New American Economy. (2021, October 28). New Americans in Gainesville. https://research.newamericaneconomy.org/report/new-americans-in-gainesville/

Ong, A. D., Cindy, S. B., Toni, L. B., & Kimberly, A. W. (2006). Psychological resilience, positive emotions, and successful adaptation to stress in later life. *Journal of Personality and Social Psychology, 91*(4), 730–749.

Sierra Niño, V. (2024). Co-creating collages to visualize interpretations about language access in North Central Florida. *Technical Communication and Social Justice, 2*(1), 44–63.

Welcoming America. (2024, February 8). Welcoming Takes Root in the South: How One Florida Community Fosters Safety and Language Access. https://welcomingamerica.org/news/how-gainesville-fosters-safety-and-language-access/

11

REVITALIZING ENDANGERED LANGUAGES THROUGH COMMUNITY-LED DESIGN

The Wikitongues Approach to Preserving Linguistic Diversity and Cultural Heritage

Tobechukwu Friday, Kristen Tcherneshoff, and Daniel Bögre Udell

Introduction

Language, a living symphony, orchestrates our shared human experience. Words, like resonant notes, carry stories, knowledge, and identity across generations. Sadly, the melody is fading. 7,000 languages are spoken or signed today, but according to UNESCO, at least 2,500 languages could disappear in 80 years—erasing centuries of cultural, historical, and ecological knowledge, and precipitating the collapse of the communities that speak them (Moseley, 2010). Language extinction is not inevitable. People lose their languages to economic exclusion, political oppression, and violence. For example, in 1892, the U.S. army general Richard Henry Pratt described forced linguistic assimilation as a "humane" alternative to genocide. "Kill the Indian," he said, "but save the man" (Park & Reisch, 2022). For nearly a century, the federal government removed Indigenous children from their families and forced them into assimilating boarding schools, where they were given English names and punished for speaking their languages. Until the late 20th century, similar policies were enforced by national governments on a global scale. Far from an accident of globalization, language extinction is an intended consequence.

Though the politics of forced assimilation have softened in recent decades—few governments still explicitly prohibit the use of minority languages—the vast majority of languages remain politically unrecognized and under-resourced, an inequity that has created immense challenges for the last speakers of endangered languages to teach the next generation. On the community level, language extinction is destabilizing. Not only does it break the bond

DOI: 10.4324/9781003469995-14

between elders and children, research shows that it reinforces generational trauma, leading to higher rates of substance abuse and suicide, impeding early and primary education, and entrenching cycles of poverty (Allam, 2020; Benson, 2004; Masters, 2021). Reversing the trend of language extinction, then, is about more than just words. It underscores cultural heritage, community wisdom, and justice.

Language revitalization is possible. With the right resources, anyone can learn their ancestral language and teach the next generation, raising new native speakers and keeping their culture alive. However, support for new revitalization projects is scarce, creating a resource bottleneck that leaves too many communities on the sidelines. Wikitongues is bridging this institutional gap with the Language Revitalization Accelerator, a funded fellowship for the leaders of new and early-stage language projects. In annual cycles, the Accelerator helps fellows identify their communities' language needs and build measurable plans to meet those needs over time. Since 2021, this program has supported 40 fellows across the Americas, Africa, Eurasia, and Australia—piloting social infrastructure for language revitalization on a global scale.

As we'll examine in the following pages, design thinking plays an essential role in language revitalization. In evaluating how design informs the cultivation of social processes, we'll find practical change-making lessons for designers and activists alike.

Applying Design Thinking to Language Revitalization

How to bring a language back from the brink? The great challenge of revitalization is patience, wrapping your head around long-term thinking. In broad strokes, our goal is to reverse the generational trend of linguistic decline—and reversing a generational trend takes a generation. As part of the Accelerator curriculum, Wikitongues encourages fellows to work with their communities to build strategic plans, identifying their short-term and long-term language needs, which can be measurably met by implementing one-year, three-year, and decades-long goals. Here, design thinking is the foundation for language revitalization, as Wikitongues fellows transform an abstract process—revitalizing a language—into a solvable problem.

Design thinking is a human-centric approach to solving problems through experimentation and iteration, with empathy for the end user. Traditionally, "end user" refers to the user of a product or tool (Chan, 2024). In our case, however, "end user" refers to the last speakers of an endangered language or the members of their culture who never had a chance to learn—collectively, the endangered language community, struggling to sustain their mother tongue. Stripped to the studs, the process of language revitalization is identifying and remediating barriers to language acquisition. By participating in this procedure, language revitalization leaders will invariably advance through five design stages: *empathizing, defining, ideating, prototyping,* and *iterating* (Lane, 2020).

In *empathizing* with the end user, language revitalization leaders work to understand the unique barriers to language acquisition in their community. Though the source of language extinction is almost always political or economic marginalization, the shape of that marginalization, and the necessary tactics for unraveling it, will change based on a community's cultural and linguistic landscape. The process begins, then, with interviewing community members about their needs, and gathering insight that shapes the language revitalization strategy. In fielding community input, the language revitalization leader helps inspire others to take ownership of the future of their mother tongue. Together, they move on to *defining* their specific language challenges, and a collective vision for their cultural future, grounded in shared priorities and values.

After defining the vision, the revitalization team shifts to *ideating*, and brainstorming solutions to their stated language challenges. From popup language classes to mother-tongue media and technology, the possible tactics for language revitalization are boundless, limited only by the community's creativity. The transition to *prototyping* marks the tangible realization of ideas as practical activities, resources, and tools, which are tested and refined through continuous feedback loops in the *iterating* phase. Over time, through iteration, the community makes measurable progress toward growing the number of fluent speakers and native speakers, restoring their language's frequent, culturally relevant use. In essence, design thinking and language revitalization do not just co-exist—they are intricately linked. Now that we have established the phases of the design process, we'll explore three case studies that illustrate its implementation: Wikitongues fellows who launched language revitalization projects in the United States, Mexico, and Benin.

Windy Goodloe: Weaving a Tapestry of Support

In 2021, Windy Goodloe joined the first cohort of Wikitongues Revitalization Accelerator fellows, representing the Black Seminole community in Brackettville, Texas. Their language, Afro-Seminole Creole, tells a transatlantic story, having emerged in the 18th century when Gullah-Geechee speakers escaped enslavement in the Southern United States and joined the Seminole Nation in what is now the U.S. state of Florida (Wittich, n.d.). There, Gullah-Geechee mingled with the Seminole languages Mikasuki and Muskogee (sometimes called "Creek"), and later with Spanish, as the

Black Seminoles traveled west to the U.S. and Mexican states of Texas and Coahuila. In time, a unique cultural identity formed (Wittich, n.d.).

The Romani linguist Dr. Ian Hancock first classified Afro-Seminole Creole (also, Black Seminole, Seminole, or ASC) after visiting Brackettville in 1976 (Tcherneshoff & Udell, 2019). By then, thanks in large part to cultural discrimination and its repression in public schools, the language was already in decline, spoken by an aging population. As of 1990, the date of the last known census of the language, there were no more than 200 native speakers (Ethnologue, 2015). Today, there are likely no more than 20 (Bögre Udell & Goodloe, 2024). Compounding the challenge of revitalization, the language is also under-resourced, with few recordings and just one abridged dictionary of about 1,000 words. (By contrast, the Oxford English Dictionary [2024] counts over 500,000 words.) Beyond that, while a majority of the Black Seminole community still lives in and around Brackettville and nearby Nacimiento, Coahuila, the remaining native speakers are scattered across the Americas (Bögre Udell & Goodloe, 2024).

Windy was among the first members of her community to champion revitalization, having been inspired to keep her language alive after participating in other aspects of cultural preservation. For example, she had supported an effort to save a historically Black Seminole cemetery, helped launch a Black Seminole museum, and was among the organizers of Seminole Days, an annual holiday celebrating Black Seminole history and culture. Reawakening Black Seminole language, it seemed, was the necessary next step for her community. When Windy first set out to revitalize Afro-Seminole Creole, she used design thinking to define their most urgent challenges: the paucity of resources, a scattered population of elder native speakers, and the COVID-19 pandemic, which erected barriers to in-person organizing. Beyond that, there was a broad lack of awareness in her community. For most Black Seminoles, Afro-Seminole Creole was irreparably lost, an impractical path to heritage conservation when compared to historic sites, cultural festivals, and traditional food. Though Windy's long-term goal was to restore mother-tongue fluency in her community—and perhaps one day, raise a new generation of native speakers—she had to start at the beginning.

Windy's first objective was to rally her community in defense of their ancestral language. To achieve that, she needed to seed a core group of language champions: adults who could both learn Afro-Seminole Creole and advocate for its use, encouraging others to get involved. Over time, these early adopters would become fluent enough to teach the language themselves. Through online organizing, phone calls, and old-fashioned word of mouth, Windy identified five adults who were eager to get started. Together, they began *ideating* to chart a measurable course forward, designing solutions to their community's language challenges. To make group learning possible at the height of the pandemic, Windy and her team partnered with Dr. Hancock to organize recurring webinars on Zoom. To expand community awareness as pandemic restrictions eased, they worked language revitalization into existing heritage efforts by hosting mother-tongue popup classes at more widely attended Seminole Days gatherings. To address the relative scarcity of language materials, they made resource creation a part of the learning process, involving students in the invention of new words for 21st-century concepts, an often necessary aspect of revitalizing an under-resourced or critically endangered language.

In three years, Windy and her team have made tremendous progress toward three measurable outcomes: resource availability, community awareness, and learner fluency. For now, they're focusing on sustaining the core team of five language champions, in order to guarantee that a proficient speaker community emerges; later, they'll adjust their metrics to track the growth of the overall learner population. (To date, their most widely attended language workshop had 20 students attend [Bögre Udell & Goodloe, 2024].) Eventually, they hope, they will be able to start measuring their success with more ambitious metrics: fluent speakers, the frequency of daily use, and finally, the number of new native speakers. In this sense, *iteration* will inform their work going forward. Though they have a long way to go before Afro-Seminole Creole is once again the predominant mother tongue of the Black Seminole community, design thinking has made it possible to structure an approach to this generational challenge. In Windy's own words, "Our dream of becoming fluent speakers again is well within reach"—for the first time in decades (Bögre Udell & Goodloe, 2024).

Hilario Poot Cahun: Seeding Voices, Harvesting Power

A thousand miles south of Windy, a member of the second Wikitongues cohort applied these participatory design methods to ensure Yucatec Maya was taught to younger generations. As a language advocate and linguist with a background in the tourism industry, Hilario Poot Cahun intimately understood the paradox of Indigenous language rights in the Yucatán Peninsula. In Quintana Roo, where he is based, two distinct worlds exist side by side: one emphasizes the historic structures of X-làabch'e'en and Chacchoben; the other, unfolding along the coastline and islands is defined by hotels, resorts, and construction sites representing Cancún and Cozumel. Simultaneously connected and yet apart, these worlds collide as the government-funded roads transport tourists from their beachfront resorts to the Maya ruins (Cruse, 2021; Yamasaki, 2020), while Yucatec Maya people venture from their hometowns to find jobs on the rapidly growing coast. Originally, Hilario was a member of the latter, with plans to continue studying English to move forward in the tourism industry. However, he quickly realized that he had the opportunity to improve the way people viewed Yucatec Maya, his mother tongue. "I learned that speaking Maya was not equal to washing dishes and cleaning rooms. I could [now] give power to the things my grandparents did, to ask people to see [the value] in Mayan knowledge" (Basalone & Poot Cahun, 2022, n.d.).

The first challenge? Reinstilling linguistic pride in his community, nowadays numbering around 800,000 people (Ethnologue, 2022). For centuries, Mayan civilization stood as a global pioneer in mathematics, engineering, and astronomy. Yet, this rich knowledge faced a precipitous decline, first at the hands of the Spanish conquest in the 1500s and later to the restrictive currents of Spanish-centric nationalism. The consequences were profound, manifesting in bans on mother tongue education for Indigenous children, a suppression that endured until as recently as 2003 (van der Haar, 2004). Even now, the confluence of low language prestige and the presence of inadequately trained and compensated teachers form a formidable challenge for the Maya community. Aspiring students find themselves forced to continue their education in Spanish, compelled to forsake their heritage language. This hierarchy, imposed by the education system, leaves many Indigenous students feeling as though their language is obsolete (Poot Cahun & Tcherneshoff, 2023).

For decades, research has shown that children receive a more comprehensive education and are able to learn faster if attending school in their mother tongue (Benson, 2004; Ozfidan, 2017; Singh, 2014). Through this language of instruction, there is less rote learning of memorization and more hands-on learning, emphasizing a truer, deeper understanding of the content and igniting creativity (Benson, 2004). These together create the ideal educational setting for intuitive learning. This form of education goes beyond propelling academic achievements: by acknowledging and affirming the cultural identity of students, it also helps enhance their self-confidence and mental well-being. In Australia's Indigenous-majority Arnhem Land region, for example, school attendance rates nearly doubled from 50% to as high as 95% after mother tongue education was incorporated into school curriculums (Masters, 2021). Some schools in the region saw, for the first time, a majority of graduates qualifying for college (Allam, 2020).

Through an education system acknowledging their heritage, children have space to affirm their Indigenous backgrounds as an enriching part of their identities, rather than hindrances (Yamasaki, 2020). In a nod to the innovative scientific creations of the Mayan civilization, and the importance of those same applications today, Hilario devised a culturally grounded STEM (science, technology, engineering, and mathematics) curriculum (Poot Cahun & Thigpen, n.d.). To start the project, he turned toward his community. He interviewed parents, teachers, and students, encouraging them to define their educational needs and goals. From this data, he created the first *prototype* of the curriculum, which was tested throughout the course of a school year in five Indigenous primary schools of Quintana Roo. This real-world application served as a litmus test, allowing Hilario to gauge its effectiveness and gather invaluable insights. Yet, true to the *iterative* nature of design, Hilario didn't stop there. After the first year, he conducted thorough surveys, once again engaging the students, parents, and teachers in the refinement process. Hilario has repeated this process three times—as of 2024, they are completing the final year of the pilot project (Poot Cahun & Tcherneshoff, 2023). Following this, Hilario and his team are carrying out an additional round of interviews and surveys before finalizing and distributing the curriculum. They anticipate that the success of their project will inspire more schools in the region to adopt the new curriculum.

Through this work, Hilario is not merely crafting a curriculum; he is cultivating a space where young minds can witness the practical application of the Yucatec Maya language. The impact is tangible: following the inaugural year of the pilot project, Hilario's interviews with students revealed a heightened sense of connection with their Mayan heritage (Poot Cahun & Tcherneshoff, 2023). Anecdotes suggest an increased enthusiasm for school among students, showcasing the project's positive impact on their overall experience (Poot Cahun & Tcherneshoff, 2023).

Ross Patrick: Nurturing Spaces From Margins to Megabytes

In 2023, Ross Patrick joined the third cohort of Wikitongues fellows, representing the Dendi people, whose homeland straddles the borders of Benin, Nigeria, Niger, and Togo. Unlike Afro-Seminole Creole, which is in critical condition, and Yucatec Mayan, which though threatened has reawakened, Dendi is in a stronger position. According to the last known census of the language, which was conducted sometime between 2000 and 2021, there are over 400,000 speakers, encompassing a majority of the ethnic Dendi population and a substantial population of non-native speakers from other cultures (Ethnologue, 2023). In other words, not only are a majority of Dendi people still native speakers, but their language is widely learned as a second (or auxiliary) language for regional trade. Beyond that, the language is officially recognized as a minority language in Benin, where a majority of Dendi live, theoretically guaranteeing a degree of institutional support. Strictly speaking, Dendi is not an endangered language—yet.

Despite the Beninese government's nominal support for Dendi, and the fact that children still learn the language, there are long-term cultural trends that drove Ross to conclude that revitalization was necessary to keep his mother tongue alive. First, the Beninese government's support for minority languages is strictly nominal, allocating few resources to support their development and daily use (Trudell & Reeder, 2006). There are few, if any, public media in Dendi, for example, and the language, though technically official, isn't taught in primary schools. Rather, French remains the sole language of public life, a vestige of colonial rule (Pomeyon, n.d.; Trudell & Reeder, 2006). In the absence of institutional support, Ross observed, daily use was on the decline. In that, he *defined* the problem. His long-term goal was securing meaningful institutional support for the language. To build a constituency for reform, he would first need to buffer against the decline of its daily use—starting with mother-tongue media. Through *ideation*, he determined that his first step was getting Dendi online.

Though it is often seen as a homogenizing force, the Internet and its related technologies can be great equalizing forces for endangered and under-resourced languages (Belmar & Glass, 2019). For example, research in Ethiopia has shown that mother-tongue access on mobile devices can be a powerful safeguard against language decline (Zaugg & Tilahun, 2019); and in Catalonia, Spain, where the growth of Catalan since 1980 is widely regarded as a language revitalization success story, media has helped drive mother-tongue development. In fact, in 1984 the regional government in Catalonia committed to subsidizing private sector media creation, guaranteeing an ecosystem of Catalan-language journalism, television, and literature, normalizing its use outside schools and public institutions. By the turn of the 21st century, there was widespread, organic demand for Catalan media which, beyond reducing the need for government subsidies, spawned grassroots campaigns for getting the language online. In 2012, for example, Catalan-speaking users translated the Twitter interface (Gomez, 2012), and in 2001, La Viquipèdia Catalana became the third language edition of Wikipedia, three months after English and German and two months *before* Spanish ("List of Wikipedias," 2023).

Today, there are upwards of 365 mother-tongue editions of Wikipedia ("List of Wikipedias," 2023), and with some exception, virtually any culture can add their language to that list. Launching a new language version of Wikipedia incorporates community organizing, translating articles from existing Wikipedias into the target language, and composing new articles in the target language. This process, Ross determined, made the online encyclopedia an ideal vehicle for digitizing Dendi and cultivating a base of activists to promote its use. To date, Ross has built a team of Dendi speakers around the cause of launching a mother-tongue Wikipedia. Through online organizing and in-person events, they have composed about 60 articles in the language and are awaiting approval from the Wikimedia Foundation, which governs Wikipedia, to move out of beta and formally launch under a Wikipedia subdomain (likely ddn.wikipedia.org, after Dendi's ISO 639-3 code). After launching, they plan to shift their success metrics to tracking the growth in articles, representing the expansion of mother-tongue resources, and measuring the number of active users, representing the impact of these resources on Dendi society.

By using design thinking, Ross was able to answer an abstract problem—the Dendi community's declining use of their own language—with a measurable, structured solution. In less than a year, he seeded the first-ever movement to sustain his language, expanded resources for mother tongue education, and got his language online. Now, as with Windy and Hilario, iteration will guide his work going forward, as his community passes their language from one generation to the next, measuring success by defining problems and testing solutions to their redress.

Getting Started with Language Revitalization: Sample Exercises

In this essay, we have explored design thinking as a change-making framework. Through *empathizing, defining, ideating, prototyping,* and *iterating,* it's possible to chart a measurable course for impact. In language revitalization, as we have discussed, this means engaging your community to identify short-term and long-term language needs and implementing those needs by meeting one-year, three-year, and decades-long goals—transforming an abstract social process into a solvable problem. The initial stage of strategic planning draws from aspects of *empathizing, defining,* and *ideating,* and can be achieved through structured discussion, community surveys, and a mix of qualitative and quantitative research. As the above case studies illustrate, specific tactics and strategies for implementing language revitalization through *prototyping* and *iterating* will change from community to community, depending on social and cultural contexts. However, the following exercises are widely applicable, driving essential aspects of language revitalization: linguistic and cultural documentation, language learning, cultural creation, and community organizing.

Linguistic Cultural Documentation

- **Community Interviews:** Organize video interviews with elders and fluent speakers, documenting cultural practices, traditional knowledge, and historical narratives.

- **Interactive Maps:** Create online maps where community members can pinpoint significant cultural landmarks and add descriptions or audio recordings in the endangered language.
- **Digital Archives:** Establish a collaborative platform where community members can contribute photos, videos, and written materials showcasing various aspects of their cultural heritage.

Language Learning

- **Sentence Construction:** Create online games where learners drag and drop words to form grammatically correct sentences, reinforcing sentence structure and word order.
- **Verb Conjugation Challenges:** Design interactive quizzes where learners conjugate verbs in various tenses and moods, practicing verb conjugations in a fun and engaging way.
- **Fill-in-the-Blanks:** Develop interactive stories or dialogues with blanks for learners to fill in using appropriate vocabulary and grammatical structures.
- **Community Tutors and Mentorship Programs:** Facilitate mentorship programs where fluent speakers guide and support learners in their language journey, fostering intergenerational connections and language transmission; also known as the *master-apprentice* model (Hinton, 2002).

Cultural Creation

- **Community Song Creation:** Facilitate workshops where learners collaborate on writing and composing songs in the endangered language, promoting cultural expression and language practice.
- **Interactive Folktales:** Develop digital folktales with branching narratives, allowing learners to make choices that influence the story's direction and practice different vocabulary and sentence structures.
- **Multilingual Poetry Projects:** Organize poetry workshops where learners create poems in both the endangered language and another dominant language, fostering intergenerational language transmission and cultural appreciation.

Community Organizing

- **Language Exchange Programs:** Partner with speakers of other endangered languages to create online exchange platforms where learners can practice conversation and cultural exchange.

Conclusion

Throughout the Accelerator pilot, the guiding principle has been community-led design, partnering with endangered language communities to co-create sustainable pathways for language preservation and documentation. By applying these design principles, we can breathe new life into endangered languages. Aspiring advocates and organizations seeking to underscore collaborative approaches should accord precedence to the following key areas:

- **Embrace community-engaged design:** Community-engaged design goes beyond consultation; it involves co-creation. Engage community members in design thinking workshops, where their input shapes the entire process.
- **Utilize participatory methods:** Participation means more than involvement; it means giving community members agency. Facilitate community-led initiatives like language festivals, where community members actively contribute to planning, organizing, and presenting linguistic and cultural content. This hands-on involvement strengthens their sense of ownership in the revitalization process.
- **Build partnerships:** Partnerships should be diverse, encompassing various fields of expertise. Collaborate with local linguists for linguistic insights, technologists for digital tools, and educators for effective teaching strategies.
- **Prioritize accessibility:** Accessibility isn't just about technology; it's about meeting people where they are. For example, develop physical language-learning kits that can be distributed within communities, ensuring accessibility for those without regular internet access.
- **Focus on capacity building:** Capacity building involves skill development for sustained impact. Train community members in language documentation, digital storytelling, and other relevant skills to ensure long-term sustainability.
- **Promote awareness and advocacy:** Advocacy is a continual effort requiring varied communication strategies. Collaborating with local media, organizing awareness events, and showcasing success stories help build a groundswell of support for language revitalization at various levels.

In its pursuit of language revitalization, Wikitongues has not only pioneered an innovative framework, but has also translated these principles into impactful projects worldwide, exemplified by the endeavors of individuals like Windy in Texas, Hilario in Mexico, and Ross in Benin. Embracing the ethos of social justice, the design principles discussed here not only provide a roadmap for effective language revitalization but also serve as a catalyst for broader societal transformation. Through this lens, we are reminded that the fight for linguistic diversity is not merely a battle for words, but a celebration of the human spirit itself.

References

Afro-seminole creole. Ethnologue. (2015). https://www.ethnologue.com/language/afs

Allam, L. (2020, December 5). Indigenous students from Bilingual School in Arnhem Land first in community to qualify for University. *The Guardian.* https://www.theguardian.com/australia-news/2020/dec/06/indigenous-students-from-bilingual-school-in-arnhem-land-first-in-community-to-qualify-for-university

Basalone, D., & Poot Cahun, H. (2022). Combining ecological knowledge in Mayan communities. Personal communication.

Belmar, G., & Glass, M. (2019). Virtual communities as breathing spaces for minority languages: Re-framing minority language use in social media. *Adeptus, 14.* https://doi.org/10.11649/a.1968

Benson, C. (2004). The importance of mother tongue-based schooling for educational quality. In *Background Paper for the Education for All Global Monitoring Report 2005: The Quality Imperative* (pp. 55). https://unesdoc.unesco.org/ark:/48223/pf0000146632

Bögre Udell, D., & Goodloe, W. (2024). Personal communication.

Chan, E. (2024, May 13). End users in UX - meaning, example, and importance. *UserBit.* https://userbit.com/content/blog/endusers-ux-terms

Cruse, S. (2021, December 15). Mexican government invests $470 million to fix tourism roads and infrastructure around Cancun. *The Cancun Sun.* https://thecancunsun.com/mexican-government-invests-470-million-to-fix-tourism-roads-and-infrastructure-around-cancun/

Dendi. Ethnologue. (2023). https://www.ethnologue.com/language/ddn

Gomez, L. (2012, July 5). Twitter now in Catalan and Ukrainian. Twitter. https://blog.twitter.com/en_us/a/2012/twitter-now-in-catalan-and-ukrainian

Hinton, L. (2002). *How to keep your language alive: A commonsense approach to one-on-one language learning.* Heyday.

Lane, L. (2020). Interstitial design processes: How design thinking and social design processes bridge theory and practice in TPC pedagogy. In M. Klein (Ed.), *Effective teaching of technical communication: Theory, practice, and application* (pp. 29–43). The WAC Clearinghouse. https://doi.org/10.37514/tpc-b.2020.1121.2.02

List of Wikipedias. (2023). https://meta.wikimedia.org/wiki/List_of_Wikipedias

Masters, E. (2021, July 5). Growing number of Aboriginal communities setting up independent schools to teach "both ways." *ABC News.* https://www.abc.net.au/news/2021-07-04/teaching-blends-aboriginal-culture-language-western-numeracy/100237208

Moseley, C. (Ed.). (2010). *Atlas of the world's languages in danger.* Third edition. UNESCO. https://doi.org/10.1086/jar.67.2.41303301

Oxford English Dictionary. (2024). https://www.oed.com/?tl=true

Ozfidan, B. (2017). Right of knowing and using mother tongue: A mixed method study. *English Language Teaching, 10*(12), 15–23. https://doi.org/10.5539/elt.v10n12p15

Park, Y., & Reisch, M. (2022). To "elevate, humanize, Christianize, Americanize": Social work, white supremacy, and the Americanization movement, 1880–1930. *Social Service Review, 96*(4), 779–835. https://doi.org/10.1086/722095

Pomeyon, G. Y. (n.d.). Statut de la langue française au Bénin: La"Revanche" des langues nationales. *Potomitan.* http://www.potomitan.info/ewop/revanche.html

Poot Cahun, H., & Tcherneshoff, K. (2023). Personal communication.

Poot Cahun, H., & Thigpen, R. (n.d.). *Kanan K'ak'náab Ma' Su'up'il* (H. Poot Cahun, Trans.). https://scholarspace.manoa.hawaii.edu/server/api/core/bitstreams/daf44453-ced8-456f-a199-530e4d5edc2f/content

Singh, P. K. (2014). Nurturing linguistic diversity in Jharkhand: Role of the mother tongue. *Economic and Political Weekly, 49*(51), 17–19. http://www.jstor.org/stable/24481144

Tcherneshoff, K. N., & Udell, D. B. (2019). Clearing space: Language reclamation, decolonization and the internet. *Book 2.0, 9*(1), 105–119. https://doi.org/10.1386/btwo_00010_1

Trudell, B., & Reeder, J. (2006). *Conference on languages and education in Africa.* SIL International. https://www.sil.org/system/files/reapdata/14/89/55/148955234815463000932035262649044414877/Trudell_and_Reeder_2006_Discourses_of_authority.pdf

van der Haar, G. (2004). The Zapatista uprising and the struggle for indigenous autonomy. *Revista Europea de Estudios Latinoamericanos y del Caribe/European Review of Latin American and Caribbean Studies, 76,* 99–108, http://www.jstor.org/stable/25676074

Wittich, K. (n.d.). *Black seminoles: A historical overview.* Seminole Nego Indian Scouts Cemetery Association. https://www.seminolecemeteryassociation.com/black-seminoles-a-historical-overview.html

Yamasaki, E. (2020). Yucatec maya language on the move: Considerations on vitality of indigenous languages in an age of globalization. In L. Siragusa & J. K. Ferguson (Eds.), *Responsibility and language practices in place* (Vol. 5, pp. 99–114). Finnish Literature Society. https://doi.org/10.2307/j.ctv199tdgh.8

Yucatec Maya. Ethnologue. (2022). https://www.ethnologue.com/25/language/yua

Zaugg, I., & Tilahun, A. (2019). *Imagining a multilingual cyberspace.* Finding CTRL. https://findingctrl.nesta.org.uk/imagining-a-multilingual-cyberspace/

12
CONTEMPORARY CHINESE GRASSROOTS ACTIVISM FOR SOCIAL JUSTICE

The Chained Woman's Case

Liping Yang and Xiaobo Wang

Social Justice in Technical and Professional Communication

Before we delve into social media activism and grassroots transnational feminist activism of the chained woman case in China, we want to share with our readers the rhetorical contexts of this particular event and how difficult of a task this has been for the netizens and citizens in China to do such work. We adopt basic concepts and issues in social justice work and technical and professional communication (TPC) from Walton et al. (2019) and others as we think it is important to bring into dialogue theories and practice of intersectional social injustice, especially the kinds of overwhelming and intersectional oppressions the Chinese citizens have to go through in order to achieve their social justice goals.

When talking about oppression, Walton et al. (2019) emphasized Young's (1990) idea of why we should first think about oppression and what exactly oppression is when we consider issues of justice or injustice. That oppression can range from physical violence to micro-aggressions to the suppression of thoughts. Ideas and lived realities within systems of domination and injustice make it important to understand that technical communication is not neutral due to technology carrying ideologies and therefore creating oppressive systems (Walton et al., 2019). This is an automatic connection with the Chinese rhetorical context being suppressive of social media users' freedom of expression on multiple social media platforms.

An important concept from the social justice framework that we want to mention from social justice theory is "marginalization." The concept is considered the most dangerous of oppressions (Walton et al., 2019, p. 19)

DOI: 10.4324/9781003469995-15

that excludes particular groups from meaningful participation in society (Young, 1990, p. 53). We believe the Chinese chained woman case has shown once again the dangerous oppression that excludes not just one or a group, but all citizens from meaningful participation in Chinese societal progress due to the heavy internet policing and overwhelming surveillance on all kinds of social media platforms. Therefore, social media users would have to find creative ways or invent their own agencies to achieve their social justice goals.

Violence is another concept relevant to social justice work. It can refer to physical and psychological attacks and/or threats as well as the reliving and witnessing of violence and violent acts across mass and social media that harm people's bodies, minds, and possessions. Applying the concept of violence to the chained woman case and the Chinese social media use context, we see an overall indifference to netizens/or users' identities and stories, and making them absolutely invisible due to the constant deletion of information and the deprivation of users' accounts. This is overlapped by another social justice concept "exploitation" (Walton et al., 2019, p. 27). The humanity of the users was denied by erasing their comments and/or accounts with the justification of exploiting their labor of posting the contents and writing their stories on the basis of violating internet regulations or posting certain sensitive words and/or phrases, which is extremely common on platforms such as Weibo and Douyin since they were made to face a public audience as compared to WeChat, which had the "Friends' Circle Feature" that enables users to share information only to those who are closer to them. Nevertheless, we do see a convergence of such features on multiple platforms as they all adopt privacy features such as sharing a post only to chosen audiences.

Powerlessness is yet another layer of social injustice in societies, which is quite relevant to the Chinese rhetorical context of doing social justice work. People experience powerlessness when they lack autonomy and authority, the ability to engage in creativity or decisions in their work/life, and status that commands respect (Young, 1990, pp. 56–57). Young gave a further example of how most people are powerless in workplaces because few of such organizations they are in are democratically organized, policies are rarely affected by public engagement, and those policies tend to be hierarchically implemented (Young, 1990, p. 56). We argue that this powerless dimension of social injustice is extremely fit in the Chinese rhetorical situation of the chained woman case, which means not only powerlessness in the workplace but also civic engagement in general. Walton et al. (2019) recommend us to center the individual and her experiences, quibbling less over definitions of power but rather committing to active listening and understanding (p. 24). We call for the understanding of the Chinese citizens' activist work and the kind of powerlessness they suffer from on all kinds of social media platforms and in reality.

To better understand issues of power and how powerlessness can lead to social injustice, Walton et al. (2019) adopted Collins's four domains of powers (p. 118). We consider the chained woman case to fit squarely within the domains of structural, hegemonic, and interpersonal powerlessness regarding social media use, human trafficking, and the liberation of the chained woman. Structurally, at the macro-level of social organization, the rhetorical situation is not optimistic here in the Chinese context. First, although the design of social media platforms such as WeChat and Weibo has enabled some activist or grassroots activism, the design of the algorithms regarding state and national policies is generally not favorable to users and their experience.

Social Media for Social Justice Combatting Surveillance on Weibo

Chen and Wang (2020) offered a nuanced picture of how feminist netizens and activists interact with changing forms of state surveillance and information control, as well as a patriarchal and nationalist internet culture and how transnational influences both support and challenge the development of productive changes in gender equality in China. Li (2020) pointed out that the commercialization of media through promoting "opinion leaders" on the Weibo platform led to an elitist discursive environment, dominated by patriarchal ideologies. Wei and Rose (2022) explored in-depth forms of resistance feminist activists in China make despite increasing crackdowns on free speech and public expression. Hou (2015) claimed that Weibo has become a new platform for feminism and grassroots activists were able to do coalition work on this platform. Yang (2016) stressed the important role of hashtags in contemporary activism such as #BlackLivesMatter.

WeChat for Activism

Deluca et al. (2016) found that WeChat was used in ENGOs' activism, transformative events of environmental activism in China. ENGO activists use WeChat to organize events and provide residents with environmental information. They believe WeChat "offers a different way to connect, communicate, organize, and disseminate information" (Deluca et al., 2016, p. 332). Wang and Gu (2016) also found that WeChat facilitates democratic design and civic engagement in transnational feminist movements, "the detainment and realization of five feminist activists" in China, and "under the dome" activism (32). It became the main channel for information dissemination before the government censored and deleted the discussion

(Deluca et al., 2016). Ma found that during the COVID-19 pandemic, WeChat was used as a platform for participatory journalism that greatly facilitated public discourse (Ma, 2021). Sun and Wright (2023) found that WeChat was proven effective when citizens challenged the censorship system and attempted to keep the "COVID Whistler" story online through "relay activism" despite government censorship. Huang (2017) also found that Chinese social media became the resource employed by different actor-speakers in political-economic ecology; among them, WeChat provides an intimate discourse community that empowers grassroots reach, mobilizes new social groups, and affords grassroots NGOs more leverage to directly engage actor-speakers from the party-state and civic society to maximize their advocated interests with instantaneity, interactivity, and a decentralized network structure (Huang, 2017). Therefore, WeChat, with its popularity and unique technological design, has been used as a key "digital network infrastructure" (Joyce, 2010), to leverage Chinese grassroots activism and promote social justice and make socio-political changes. Our chapter mainly focuses on the multimodal and multilevel activism that was enabled by the features on WeChat, as Douyin, Weibo, and other platforms simply cannot support such activism. However, we still present the struggles of users on Douyin and Weibo as such platforms still offer affordances to some extent within a limited time and through the creativity and resilience of activist users.

Rhetorical Analysis

We use frameworks such as digital activism on Weibo and WeChat, social justice in TPC (Walton et al., 2019), Transnational Feminism on social media (Chen & Wang, 2022, 2023), and fluid digital community engagement (Hinck, 2017). These frameworks help us better understand and analyze complex cases such as the chained woman in China, transnational feminist activism, digital discourse flow, evolution, and consequentiality of grassroots social activism. We specifically look at how different forms of media, multimodal text, and narrative styles influenced the dissemination and reception of information related to the case.

Methodologically, we analyze the written language, audio or video text, and visual narratives used in WeChat posts and comments. Additionally, we construct a detailed timeline of key events, hashtag trends, and policy and news announcements to understand the dynamics of the case over time. Our findings reveal a significant shift in public opinion and policy, driven largely by grassroots activism. This study not only sheds light on the chained woman case but also contributes to a broader understanding of the role of WeChat in shaping social justice narratives in the digital era.

Rhetorical Strategies in the Chained Women Case

Our research found users/citizens fighting against social injustice during the development of the chained woman case via the use of rhetorical strategies such as visual remix and storytelling, as well as community building on the Chinese social media platform WeChat. We found such strategies to be conducive to the exposure of the chained woman case in the first place, grassroot activism to combat powerlessness and intersectional oppression for social justice. However, we decided to focus on WeChat instead of Weibo and Douyin due to the heavy policing, and erasing of information and accounts, as well as the more public nature of the two platforms.

According to the latest Global Digital Report, Kiely (2023) reported that 90.6% of Chinese people using the internet do so via mobile devices and platforms like Weibo (the most popular Chinese social media apps, with more than 605 million monthly active users), WeChat (1.309 billion monthly active users, Shewale, 2024) and Douyin (the TikTok platform based in China that ranks second on the most popular Chinese social media according to Statistia, 2024) for daily communications, news, and other needs.

Short Video on Douyin Exposed the Chained Woman

Douyin's status in China has been extremely popular thanks to its "short video" branding/feature that offers users options to post videos from 3 seconds to 10 minutes long. Since its launch back in September 2016, Douyin has gained global popularity (both Chinese *douyin* and global TikTok platforms), with 746.5 million monthly active users as of 2023 on Chinese Douyin (Verot, 2023) and the global TikTok boasts over 1 billion monthly active users and more than 4.1 billion times downloads (Backlinko Team, 2024) as of 2024.

The chained woman was first exposed to the public on the evening of January 27, 2022, when a Douyin user "Xuzhou First Repair Bro" posted his short video visiting Zhimin Dong, another Douyin user Zhimin Dong, who had posted about his eight kids and the hardships he was going through in the hope of collecting public donations for the living of his kids. Due to our IP addresses, we cannot access the Douyin video that was originally published, but here is a screenshot of part of the original video from YouTube (Yunfei Official, 2022), which was a repost and combination of other videos that were posted after the exposure of this case to the public. The original Douyin video has been deleted or erased from the Douyin platform, which we have no way of knowing. We also have no way to know which user was the one who posted the original chained woman video.

With the help of the hashtag "the chained woman," the video went viral and was reposted to different social media platforms such as Weibo and WeChat. Her experience of being abused, being human trafficked, and accusation of having mental illness shocked the entire country, leading many citizens, journalists, and lawyers to support her at Xuzhou County ("The Internet Has Memory, Don't Forget the Chained Woman"). According to the same source, up until March of 2022, the most searched three hashtags about the chained woman on Chinese Weibo went beyond 10 billion times/views, which was comparable to hashtags about the Beijing Olympics Games that were officially advertised. Many netizens had stepped out bravely to share their own stories about their mothers, daughters, sisters, and classmates being kidnapped and disappeared thereafter, which led to more discussions on social media about human trafficking, women's rights, sexual abuse, rape, family violence, how to protect people with mental health, and so on. Although the video is long gone from Douyin and other platforms in China, we were able to access part of it through YouTube and we hereby offer a summary of what happened in the short video:

- Douyin user "Xuzhou First Repair Bro" visited Zhimin Dong, the "eight kids' father" who had gained internet fame by posting his hardship raising the kids and therefore asking for donations from the public;
- The shocking visuals and poor living conditions of the chained woman were revealed first time ever the Chinese public;
- "Xuzhou First Repair Bro" asked the chained woman several questions and invited her to tell the audience how she felt about her situation.

Figure 12.1 shows the only two accounts that are named "Xuzhou First Repair Bro" and neither account is showing any chained woman video. These two accounts seemed not even the original "Xuzhou First Repair Bro" and the accounts may have been created to attract more views. We

FIGURE 12.1 Screenshots of "Xuzhou First Repair Bro" on Douyin.

also present here Figure 12.2, a screenshot from a YouTube video that is part of the original Douyin video when the chained woman was first discovered. The figure shows the chained woman wearing a coat "Xuzhou First Repair Bro" gave her when he saw her in summer clothing. The caption reads "Like a prostitute" as she was asked about how she was doing.

FIGURE 12.2   Screenshot of the first Douyin video. Caption in screenshot reads: "Like a prostitute."

All the keywords such as "the chained woman", "xuzhou first repair bro", "xuzhou eight kids mother" from this Douyin video became hashtags that users can spread the information, tell the story, and make connections to take action.

In the YouTube video, the chained woman told user "Xuzhou First Repair Bro" that "the world has ditched me" and "I feel like a prostitute." She was chained to the corner of a shabby room and wore only thin summer clothing in winter times when the video was shot. On the floor, there was a bowl to feed her. All of her eight kids were living in another room that seemed much more decent with Zhimin Dong. The shabby room was a side house in the courtyard of the Dong family.

According to "The Internet Has Memory, Don't Forget the Chained Woman" (2022), up until March of 2022, the most searched three hashtags about the chained woman on Chinese Weibo went beyond 10 billion times/views, which was comparable to hashtags about the Beijing Olympics Games that was officially advertised. Many netizens had stepped out bravely, especially on Weibo and in real life, to share their own stories about their mothers, daughters, sisters, and classmates being kidnapped and disappeared thereafter, which led to more discussions on social media about human trafficking, women's rights, sexual abuse, rape, family violence, how to protect people with mental health, and so on.

Activism on Weibo to Support the Chained Woman

Weibo, as the largest public social media platform widely used among Chinese diaspora populations, has played an important role in mobilizing public opinions and enabling social justice and grassroot activism during the chained woman grassroots activism. Weibo users go to "hot searches" for the most discussed events and topics, and they use the hashtag feature primarily when engaging their audience. However, due to its public nature in the first place, Weibo undergoes censorship the most. Hashtags, sarcasm, video/visual sharing, and direct questioning of authorities have been the primary approaches to build the activist community during the chained woman case (Figure 12.3).

User "韦观Leo" published a short post with a link to the original video report using hashtag #OfficalReportontheWomanWhoGaveBirth-toEightKidsatFengCountyXuzhouProvince (English translation), praising Shenzhen Satellite Television on being the first television station/channel that had reported the chained woman case officially with more details from the exposure of the case to the public to an investigation team was formed. User "何时见陆", an influencer who is a verified user, said, "....I

FIGURE 12.3 Hashtags, video/visual sharing, storytelling, sarcasm, and direct questioning in Weibo activism.

am raging, I am voicing my opinions, but they said I'm doing boxing… If I stop checking on this matter, these women will really be forgotten, and there will never ever be a chance for them to be saved." According to Chen and Wang (2022, 2023), the Chinese feminists and activists had to contribute extra emotional labor and go beyond the features of the social media apps than what is generally expected. Policies largely lead to an overall surveillance and internet policing culture that is extremely hard for users to enjoy the freedom of the information age. We can see from the posts and comments shared here that users rely on sarcasm, perseverance, and often blunt questioning to harness their rhetorical agency. User "奉还还" complained in a sarcastic way, "If you delete all the information, you will then have solved all the problems." User "Coeurcc" complained, "Some people commented under my Weibo post 'Has the case been reported and resolved?' They might have only seen the 'reported' part and forgot about a real solution," which blames the inaction of different levels of governments in a sarcastic tone.

By using hashtags, users can easily find the chained woman case and the numerous videos, visuals, and stories other users have posted. Using sarcasm and direct questioning, users were able to voice their opinions against the heavy censorship and demand transparency in social investigations and policy making. Many of the comments were complaining about the inability of different levels of government, and corruption that prevented the case from being truly investated and therefore giving the public true details of what happened, thus forming a community where netizens unite together to free themselves via the liberation of this one woman.

Multidimensional Activism on WeChat in Support of the Chained Woman

WeChat has played a crucial role in the case of the chained woman activism through its capability to promote self-publication and journalism via WeChat Official Accounts (OAs), facilitate free speech and divergent discourse flow, and use visual remixing and multimodal storytelling to create relay activism through community bonding, further avoid China's Internet censorship and government surveillance, form counternarratives and public opinion to compel government for deeper investigation, enhance user privacy and create stronger community bonds.

Self-publishing and Journalism Through WeChat's Official Accounts

WeChat's OAs are a significant feature that enhances activism by increasing visibility and popularity. Unlike other social media platforms in China, WeChat's OAs function supports individuals or organizations in setting up OAs for informal self-publication. OA authors can publish and circulate articles to the public audience through subscription, general search, reposting, and sharing. This functionality is distinguished by its ability to support relatively lengthy content, free from the constraints of a single modality. In this sense, WeChat OA serves as a tool for self-publication, facilitating organized grassroots activism with the principles of free public speech and the "free flow of information" (Wang & Gu, 2016).

With OA accounts, self-publication doesn't necessarily require official review and circumvents government censorship within a short period, making it ideal for free speech. In a political landscape where social media platforms are influenced by mainstream media that don't efficiently update the true information or even mislead the public, OA motivates activists to engage in civic journalism and self-publication to disseminate information and enable social activism (Huang, 2017). With the capability of running one's own media and news outlet and being able to voice publicly one's experiences, observations, analyses, and opinions for activism., such OA self-journalism greatly facilitates information generation, flow, and dissemination. With this affordance, not only can WeChat users stay informed on a mobile contact app they use the most, but the wider public can also engage in conversations about events and even interact and challenge institutional media through public opinion formed by civic journalism. Through the opportunity for self-journalism via individual public media outlets, WeChat users become active designers who shape, redesign, and localize available technology to fit into their own local context (Sun & Wright, 2023). Therefore, individuals are no longer passive audiences but active participants in activism, voicing their experiences and opinions directly and freely.

In the case of chained woman activism, the initial exposure of the chained woman online sparked public discussion and a keen interest in further investigation. However, the local government's initial reports to the investigation into Xiao Huamei's circumstances offered wrong or misleading information about her identity, presenting a generalized narrative that the family was happy and normal, attributing the situation solely to the chained woman's mental health issues. On January 28, 2022, Fengxian County issued its first official statement, asserting that Mrs. Yang, a mother of eight, was married and her situation was not related to human trafficking. The statement also described Mrs. Yang as suffering from mental issues, known for mistreating her children and elders, and that she was undergoing treatment while the family received community support. This narrative contrasted with the initial outcry. The general public was irritated by the outrageous misinformation from the public accounts, prompting many to start their investigations, with some locals sharing their findings online. The online community, noting her accent reminiscent of the Yunnan and Sichuan provinces, doubted the claim that she was a local resident with no ties to human trafficking, as the public reports said. One user even uncovered their "marriage certificate," highlighting discrepancies with Little Huamei's travel history. Overnight, self-assigned detectives posted on various social media platforms pointing out the suspicious reports and demanding justice.

Faced with these online challenges, the Xuzhou authorities quickly issued a statement two days later. They clarified that the "chained woman" was not a local but had been brought in by Mr. Dong, who had also mistreated Xiao Huamei (Figure 12.4). This follow-up notice completely contradicted the initial report, exposing the mistreatment of the chained woman and potential human trafficking.

WeChat's self-publishing capabilities, free from political or mainstream media interference, empower individuals to share their thoughts and experiences, offering a platform for grassroots voices that are often lacking in traditional media publication channels. It functions like a network of interconnected yet independently operated newspapers, allowing both individuals and organizations to express their views in a shared media space. Regardless of their status—be it students, journalists, activists, or media influencers (users with greater popularity who had more influence among the public). This dynamic provides users the opportunity to surpass mainstream media in terms of voice and reach. Through the creation and management of their own "official" public media accounts, WeChat's self-publishing feature facilitates a higher degree of freedom of expression and a broader dissemination of ideas, enabling users to take on roles akin to

FIGURE 12.4   Fengxian News: Official reports on WeChat about the Chained Woman Case.

journalists or critics. This contributes to a media landscape that is more diverse and inclusive.

Divergent Discourse Flow

WeChat's media architecture enables a diverse flow and distribution of discourse through its public media platforms, effectively spreading information and shaping public opinion via direct self-publishing and widespread sharing and/or reposting. Vertically, the discourse flow can be likened to a "cascading network activation model," a concept from cognitive psychology that explains how information travels within the human brain. Central media agents distribute ideas from primary hubs via subscriptions, reaching an expansive audience and swaying public sentiment (Entman, 2023). Horizontally, popular articles are amplified through reposts and shares across "moments," "group private chats," "viewings," and other features, allowing for the widespread dissemination of activism-related content such as summaries, documentation, and critical analyses through both sharing and subscriptions.

Besides the discourse flow, multiple modalities of that information also contribute to the effectiveness of activism rhetoric in WeChat. WeChat's multimodal platform distinguishes its official articles as a unique genre of text production, unlike traditional publication businesses. This sets it apart from other social media platforms that may be limited by character restrictions (such as Little Red Book or Weibo) or dominated by video or image content. WeChat's multimodal articles effectively integrate various sources, ensuring efficient message distribution and content richness. They address the complexity of reporting incidents by combining narrative text, images, videos, and inviting engagement through comments. This makes WeChat's official articles a central platform for activism. Moreover, their mobile-friendly format makes these multimodal articles easily accessible for individuals who may not have the time or resources to independently research incidents.

In the grassroots activism case of the chained woman, the sheer volume of information available online about Xiao Huamei—ranging from the original exposé and platform reports to institutional notices and independent journalism on various social media—can overwhelm the audience. Institutional reports on the case are scarce and often unreliable, making it challenging for those trying to stay informed, especially if they are not closely tracking the incident, as new information surfaces daily.

Given this overload of disparate data, activists employ WeChat OAs to clarify and aggregate event information journalistically. Observing the unique offerings of other social media platforms—TikTok's focus on video content, Weibo's 140-character limit for textual criticism and sentiment, and Little Red Book's emphasis on visual appeal and descriptive analysis—activists using WeChat OA amalgamate these elements. They integrate content, effectively incorporating necessary audio and video evidence, to provide the most comprehensive and multimodal perspective on the event and its context. This approach keeps users informed with detailed articles, literary expressions of the events, socio-political analyses, critiques, and storytelling, among others.

For example, some influential account holders meticulously gather various sources, outlining the incident's evolution and sequence of events, and reinforcing their narratives with evidence like screenshots from the original TikTok live or quotes from official media posts. These articles organize information with precision, leveraging "evidence" from different social media platforms to shed light on the incident, a method unmatched by mainstream media or institutional announcements. Through OA subscriptions, sharing on Moments, and group chat reposts, these articles effectively educate the public on the issue.

Therefore, WeChat OA serves not only to educate the public about the issue and impart legal knowledge but also to engage them in discussion and capture their attention. As audiences access these educational and expressive updates on their most-used app, OA articles also act as a catalyst, motivating them to participate actively in activism.

Multimodality and Censorship

Besides integrating various content genres from different media, WeChat's multimodal capabilities support the free flow of information through diverse visual and remixed formats, essential for activism. A distinguishing feature of Chinese social media campaigns, setting them apart from US movements like #MeToo, #BlackLivesMatter, and #StopAsianHate, is the stringent internet surveillance by the Chinese Communist Party (CCP). Chinese activists disseminating information and creating content must ingeniously utilize social media's features to bypass censorship. By innovatively converting text into images and embedding text in photos, activists have successfully circumvented AI censorship, evading keyword filters (Rauchfleisch & Schäfer, 2015). This visual remixing adds a new layer to activism within the surveillance society of China. Furthermore, WeChat's audio messaging feature has become a key tool in activism, enabling the circumvention of censorship. Audio content, being more challenging to censor and enriched with local dialects or speech patterns, allows activists to convey sensitive information while keeping the discourse alive online.

In the activism surrounding the chained woman, many OA articles were swiftly deleted after publication due to government surveillance and AI censorship targeting specific keywords and content, mirroring the fate of other activism-related information in China, such as the "COVID whistleblower" during the pandemic. The government aimed to prevent the widespread dissemination of activism content, seeking to mitigate public fear of the pandemic or the negative impact of "extreme" discourse. However, some articles evaded this "content filter" by incorporating visual remixes and audio messages. By leveraging these methods, activists empowered themselves and the community, maintaining the momentum of the story through relay activism (Sun & Wright, 2023).

Similar to social media activism in other countries, campaigns in China fully exploit the potential and features of various social platforms to secure public recognition, encourage widespread participation, and achieve greater visibility, virality, and resonance. However, activists face significant challenges in maintaining their presence and momentum under the watchful eye of strict political surveillance and censorship. With WeChat OAs and its capabilities for audio narratives, visual remixing, and intimate discourse sharing, grassroots activists have been able to leverage these features to circumvent censorship to some extent. As the activism gained strength, it sustained the narrative and rallied the public to unite and advocate for women's and children's rights, as well as to uncover and combat human trafficking.

Form Counternarratives to Compel Government for Social Justice

Independent media agents on WeChat not only function autonomously but also actively contest institutional and governmental narratives. Many OA operators directly confront and question institutions in their publications, engaging the public and influencing public sentiment. This public pressure often forces the government to undertake further investigations, ultimately leading to the revelation of the truth.

For instance, critical analyses in popular WeChat OA articles often combine partial factual narratives with expressive sentiment, drawing parallels to other cases of human trafficking or abuse of women in history or different locations. These articles advocate for heightened awareness and demand institutional action. Following the Xiao Huamei incident, and before the institutions provided accurate updates or even issued incorrect statements, numerous WeChat articles scrutinized the OAs, ultimately aiding in uncovering the truth and delivering justice to the public.

Before these four institutional updates (Figure 12.5) were released, many OA authors had already been challenging the institutions for a deeper investigation and truth rather than superficial and misleading information. As more information surfaced from either institutions or private accounts, influential media account holders critiqued the underlying issues, including the reliance on social connections over law and justice, traditional societal constraints on women, and government inaction. They criticized the government's attempts to cover up the incident and examined traditional values by revealing similar incidents across China. Besides challenging mainstream and institutional media for transparency

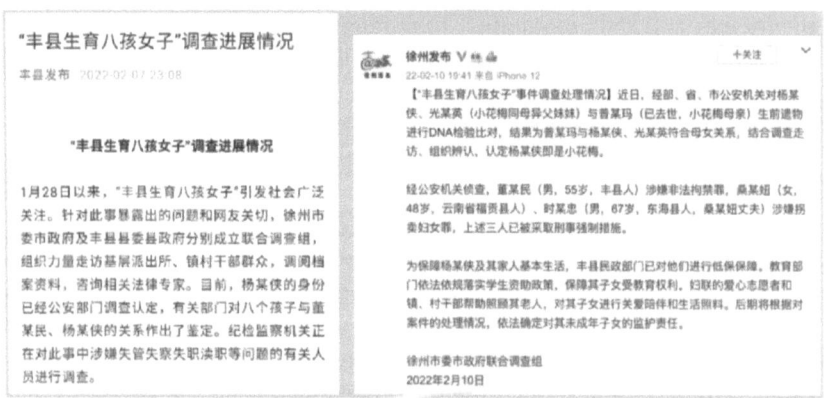

FIGURE 12.5   Progress report on the investigation of the Chained Woman incident (Fengxian Official News account).

and justice for Xiao Huamei, these critics also urged the public to deeply consider the fundamental issues behind such incidents. These critical articles acted as a catalyst, engaging public attention and collectively challenging institutional responses to ensure justice.

It is troubling to see how the government initially glossed over the case investigation irresponsibly and attempted to deceive the public with falsehoods and vague statements. Without public discourse and opinion, institutional media might even be reluctant to reveal the truth. The pressure mounted from public writings, catalytic sentiment, and criticism, urging the government toward a more thorough investigation. In response, institutions such as the county, city, police department, and court departments issued four successive updates, disseminating information through social media, OA articles, and other media channels.

Enhanced User Privacy and Stronger Community Bond

The last notable feature of WeChat's technological affordance is its privacy settings, which, in turn, foster stronger bonds and connections among individuals within the activist community. Unlike other social media platforms with open and public settings, such as Weibo, where your posts are mostly visible to the public and anyone can search, see, and comment on your posts, WeChat features "friends' circles," where most communication occurs within intimate circles or *pengyouquan, thus* enabling users to establish and maintain control over their discourse in social media. When participating in public writing and activism, users maintain control over their discourse within a trusted community through exclusive information sharing. With their privacy protected while engaging in social campaigns, users have a heightened sense of security (Wang & Gu, 2016), allowing them to post information more freely and fostering a stronger sense of discourse community. Therefore, WeChat's relative privacy seems to "undergird a more permissive space for political talk" (Tu, 2016).

In WeChat's design, the boundary between private and public is deliberately blurred (Wang & Gu, 2016). In other words, users can "publicly" share information within the "private" community. Without worrying about their words being exposed to the public or censored by the government, individuals engaging in activism can join conversations more persuasively, speak up with more enthusiasm and motivation, and are more likely to engage in community discussions. Deluca et al. (2016) found that with a stronger sense of community, people are more likely to translate online discussions into offline action and local events. Through gaining rhetorical significance from social media discussions and eventually achieving rhetorical transformation, activism on WeChat spurred the "consequentiality they spark in the world" (Gries, 2015, p. 3). Within a trusted and secure online environment, activists utilize discourse proliferation through generation, reposting, and sharing to enable a viral economy, intensifying participation with each new encounter.

In the case of the "chained woman," after the event was revealed, people attempted to contact the "women's union" for more updates and investigation and also urged the organization to pay attention to similar cases involving women trafficking, family violence, and abuse, ultimately challenging and improving community-based mental health services in China (Bai et al., 2022). Therefore, WeChat features prominently in the realm of user privacy, yet it also boasts numerous public accounts that serve as ideal platforms for community building and storytelling. Given WeChat's extensive user base and high frequency of daily use, official articles on the platform play a significant role in activism by educating the public, changing opinions, and raising awareness on various issues.

WeChat empowers independent media agents with its self-publishing capabilities and a multimodal approach, nurturing an environment for online storytelling and grassroots community building. In the case of the chained woman activism, WeChat has been instrumental in facilitating the free flow of information, bypassing censorship, and effectively mobilizing public participation and discourse, thereby raising awareness on critical issues such as human trafficking. This platform grants activists increased autonomy to share their experiences and insights while also pressing institutional authorities to address relevant concerns. Moreover, it encourages deeper reflection on issues such as women's abuse, human trafficking, and family violence, thus playing a vital role in informing and educating its users.

Conclusion

In this chapter, we have highlighted the transformative power of social media platforms in activism for social justice, specifically within the Chinese and Chinese diaspora context in the case of the chained woman case that was brought to the public's attention in January 2022 and lasted until unfair court decisions in 2023, with many contemplating the unfair court rulings against human trafficking and sexual abuse.

Our study underscores the potential and unique social justice design-driven digital platforms to amplify grassroots voices, mobilize public opinion, and policy-making on human trafficking women's and children's rights, thereby playing a crucial role in the contemporary social justice landscape of social activism. Examining popular Chinese social media platforms such as WeChat, Douyin, and Weibo, we find that their design features and affordances, like hashtags, "hot search," reposting and sharing to Moments and private group chat, and subscriptions, all facilitate both horizontal and vertical discourse flow, enabling activism messages to get viral and reach a wide audience. These platforms' self-publishing and user voice features allow grassroots users to share firsthand evidence, voice their opinions, and, most importantly, form counternarratives to challenge OAs in the same online space, drawing attention to and addressing injustice. Additionally, multimodal posting and publishing, particularly visual remixes and audio narratives, effectively help activism stories survive government filters and AI censorship, fostering relay activism through community bonding.

Features and rhetorical strategies of Chinese social media platforms and users' resilient and perseverant activism are analyzed to help better understand users' activism throughout the case, providing useful insights to future and transnational grassroots movements for social justice, as well as the teaching of design for social changes. In the design of the TPC class, observing how social media affordances can facilitate or restrict activism discourse and social justice work in online public opinion and discourse is crucial. It is also important to examine how online activism can creatively use digital technology to achieve more compared to offline activism. The pedagogical takeaways are to emphasize the rhetorical analysis of the design and affordances of social media and other communication technologies in the context of technical communication. Additionally, after case studies, the pedagogy could aim to encourage students to critically design communication technologies for social justice in their future workplaces.

We felt encouraged to see tighter legislation and call for action here for the collaborative efforts among activists, educators, technical communicators, and content designers from China and around the globe to translate, archive, and design better platforms for social justice work under heavily policed rhetorical contexts. Meanwhile, we also praise the wisdom and effort of activists and the rhetorical features of existing platforms that have enabled productive activism.

References

Backlinko Team (2024, July 1st). TikTok statistics you need to know. *Backlinko*. https://backlinko.com/tiktok-users

Bai, W., Li, X. H., Ma, Y., Su, Z., Ng, C. H., & Xiang, Y. T. (2022). Mentally ill woman chained in shack reflects the challenges of community-based mental health services in China, *Asian Journal of Psychiatry*, 73, 103154, https://repository.um.edu.mo/handle/10692/99203

Chen, C., & Wang, X. (2020). #Metoo in China: Affordances and constraints of social media platforms. In J. Jones, & M. Trice (Eds.), *Platforms, protests, and the challenge of networked democracy* (pp. 253–269). Palgrave McMillan.

Chen, C., & Wang, X. (2022). Contemporary Chinese feminist rhetorics: #Me-Too in China. *Enculturation: A Journal of Rhetoric, Writing, and Culture, 33.* https://enculturation.net/contemporary_chinese_feminist

Chen, C., & Wang, X. (2023). Reporting online aggression: A transnational comparative interface analysis of Sina Weibo and Twitter. *Technical Communication, 70*(4), 42–59.

Deluca, K. M., Brunner, E., & Sun, Y. (2016). Weibo, WeChat, and the transformative events of environmental activism on China's wild public screens, *International Journal of Communication, 10,* 321–339, https://ijoc.org/index.php/ijoc/article/viewFile/3841/1539

Entman, R. M. (2023). Cascading activation: Contesting the White House's frame after 9/11. *Political Communication, 20*(4), 415–432.

Gries, L. (2015). *Still life with rhetoric: A new materialist approach for visual rhetorics.* University Press of Colorado.

Hinck, A. (2017). Fluidity in a digital world: Choice, communities, and public values. In A. Hess, & A. Davisson (Eds.), *Theorizing digital rhetoric* (pp. 98–111). Routledge.

Hou, H. L. (2015). On fire in Weibo: Feminist online activism in China. *Economic and Political Weekly, 50*(17), 79–85. http://www.jstor.org/stable/24481829.

Huang, D. (2017). Social media and activism of grassroots NGOs in China. In M. Kent, K. Ellis, & J. Xu (Eds.), *Chinese social media: social, cultural, and political implications* (pp. 22–40). Routledge.

Joyce, M. C. (2010). *Digital activism decoded: The new mechanics of change.* IDEA.

Kiely, T. (2023, Dec. 12). The 8 top Chinese social media apps, sites & platforms 2024. https://www.meltwater.com/en/blog/top-chinese-social-media-apps-sites

Li, S. (2020). 微博女权的前世今生：从"政治正确"到"商业正确" ("The previous and present life of Weibo feminisms: From 'political correctness' to 'commercial correctness'"). *澎湃新闻, The Paper.* https://m.thepaper.cn/newsDetail_forward_7854160?from=timeline&isappinstalled=0

Ma, R. (2021). Graphic activism on WeChat: The aesthetics of dissent in the age of COVID-19. *Design and Culture, 13*(1), 33–42.

Rauchfleisch, A., & Schäfer, M. S. (2015). Multiple public spheres of Weibo: A typology of forms and potentials of online public spheres in China. *Information, Communication & Society, 18*(2), 139–155.

Shewale, R.. (2024). 18+ WeChat Statistics For 2024 (Users, Revenue & More). https://www.demandsage.com/wechat-statistics/#:~:text=WeChat%20is%20the%205th%20most,networks%20by%20monthly%20active%20users

Sun, Y., & Wright, S. (2023). Relay activism and the flows of contentious publicness on WeChat: A case study of COVID-19 in China. *Information, Communication & Society, 27*(2), 257–277.

Tu, F. (2016). WeChat and civil society in China. *Communication and the Public, 1*(3), 343–350. https://doi.org/10.1177/2057047316667518

Verot, O. (2023, September 8). Douyin statistics and trends. *Marketing to China.* https://marketingtochina.com/douyin-statistics-and-trends/

Walton, R., Moore, K. R., & Jones, N. N. (2019). *Technical communication after the social justice turn: Building coalitions for action.* Routledge.

Wang, X., & Gu, B. (2016). The communication design of WeChat: Ideological as well as technical aspects of social media. *Communication Design Quarterly Review, 4*(1), 23–35.

Wei, X. A., & Rose, K. (Eds.). (2022). *Weibo feminism: Expression, activism, and social media in China*. Bloomsbury Academic.

Yang, G. (2016). Narrative agency in hashtag activism: The case of# BlackLives-Matter. *Media and Communication*, 4(4), 13–17. https://doi.org/10.17645/mac.v4i4.692

Young, I. M. (1990). *Justice and the politics of difference*. Princeton University Press.

Yunfei Official. (2022, March 1st). *Complete video of the Xuzhou chained woman who had eight kids* [Video]. YouTube. https://youtu.be/Bij60AxT7ig?si=Ekd15Jsyp8iI9LYf

PART III

Pedagogical Exemplars of Multimodal Design for Social Justice

13

A KAIROTIC APPROACH TO TEACHING ONLINE ASYNCHRONOUS COMMUNITY-ENGAGED TECHNICAL COMMUNICATION COURSES

Antonio Byrd

Introduction

Technical and professional communication (TPC) courses often combine industry practice with community-engagement for a more well-rounded education for students; the service-learning approach teaches students that their skills and competencies gained from TPC courses well-position them to be critical citizens. The social justice turn in the discipline lends another layer of purpose: from problem-solving in general (Johnson-Eilola & Selber, 2013) to problem-solving for social inequities (Agboka & Dorpenyo, 2022b; Jones, 2016; Tham, 2021; Walton & Agboka, 2021; Walton et al., 2019) specifically. The mix of community engagement with social justice creates new potential pedagogies for students to understand their role as technical communicators in constructing better "social futures" for local and global communities (New London Group, 1996, p. 65). Technical communication and service-learning have contributed to teaching civic engagement although not always at the intersection of social justice. While civic engagement addresses issues of public concern (Harrell, 2020), those projects may not address correcting oppression against marginalized people or other forms of substantial change (Grabill, 2004).

Agboka and Dorpenyo (2022a) report on a survey study of 231 TPC programs and found that theorizing social justice in scholarship may outpace programmatic and curricula adoption of social justice frameworks in teaching. Of the 27 instructors teaching social justice, seven taught courses online, and seven taught social justice courses in a hybrid;

DOI: 10.4324/9781003469995-17

about 81 percent of respondents taught these courses in-person (Agboka & Dorpenyo, 2022a). TPC has a wide-ranging and robust commitment to online teaching (Hewett & Bourelle, 2017). Over the last several years, there's no shortage of teaching articles on facilitating community-engaged projects through online spaces (Bourelle, 2014; Cleary, 2021; Dumlao, 2023; Francis, 2018; Nielsen, 2016). Some instructors have written about deploying the 3Ps framework in online courses, a recognized and now foundational theory to social justice practice in TPC (Zamparutti, 2022). Although the number of instructors is small given the number of programs represented, Agboka and Dorpenyo's (2022a) work leads to a generative observation: Online TPC instruction should pick up the pace of adopting social justice pedagogy in a time when demand for online course offerings among adult undergraduates continue to rise (Garrett et al., 2023).

While many TPC programs take responsibility for the professional and civic lives of thousands of students going into the workforce, the field cannot overlook those instructors who teach for smaller less-known institutions that offer only one or two technical communication courses. For these instructors, as Bourelle (2014) notes, time and resources that support community-engaged and justice-informed activities outside the classroom may not be as robust. In addition, many students may work full-time jobs or have family obligations which would limit their engagement with community organizations. Sometimes an online asynchronous environment allows more flexibility to meet these life responsibilities in an already-packed schedule. However, the addition of finding one's own organization or being assigned to work with an organization may hamper that flexibility. These student populations may not be consistently represented in scholarship on teaching and learning in TPC. Writing Studies scholarship often focuses on research-intensive universities with tenure-track faculty and their students. In many examples, the specific demographics of students aren't even described. But millions more undergraduates in the United States take first-year composition courses at two-year colleges. What about these "other students"?

> ...the ones who don't volunteer for research projects, or the ones who don't attend class regularly. Our students are the ones who work forty hours a week or the one who have gaps in their knowledge education that affect classroom engagement and success. They are the writers whose confidence has often been battered by negative educational experiences and have little faith in the educational system – or teachers – to support them [!]
>
> *(p. 39)*

Despite calling for survey participants across the most well-known and well-used listservs in TPC, Agboka and Dorpenyo (2022a) write that out of 52 TPC instructors, 42 or 80.77% taught at public colleges or universities but "none of these respondents were from community colleges, vocational colleges or universities, women's colleges or universities, HBCUs, or TCU's" (2022a, p. 47). The scholars are familiar with work happening at HBCUs but that's not represented in the data. There are many reasons why these institutions don't appear on the survey. For me, the first question of the survey blocks my participation; it asks for the name or title of my program. Although I teach three TPC-related courses, I cannot answer this question because there is no such program at my institution.

The eight-week project I describe in this chapter concerns instructors teaching one or a few TPC courses in departments that must contend with students who don't have the time for significant collaboration with community partners and vice versa because they work full-time or take care of family. Kairotic pedagogy uses more strategically timed interactions. In my course, that's small, short bursts of engagement between students, community partners, and industry professionals. Greater attention to *kairos* allows instructors to be more strategic about when to consult with community partners using the design thinking mindset and Agile framework. This chapter implies that teachers outside of formal TPC programs with less privileged students can still deliver on goals and outcomes without always deeply embedding students in community-engaged work in an online asynchronous course.

First, I explain the institutional and departmental context for my kairotic pedagogy in the online asynchronous TPC course called Technical Communication. I later describe the rhetorical theory that guides my approach to designing for community engagement in TPC. Third I describe course design and focal activities that have students study social justice practices in preparation for working on an 8-week prototyping project with an optional community partner. Finally, I share results from a focus group interview that shows the ways students have reconceptualized audience, writing, and academic writing.

Context and Background

I teach at a Research II institution that serves 16,000 students in a city that has a lively nonprofit and professional sector. My institution does not have a formal program in TPC. Because local community colleges offer technical writing courses for first-year students, my institution allows transfer students to count their technical writing courses as credit, replacing the second sequence of first-year writing courses. STEM faculty have proposed that technical writing replace the second sequence of first-year writing, as well. Two challenges to this proposal persist: First, a lack of funds to hire and train what would most likely be a workforce of contingent faculty and graduate students. Second, technical writing does not teach the research methods students need for college writing; if a technical writing course appeals to a specific major, that student may change majors having none of the research methods learned from first-year writing.

As the sole TPC instructor, I teach three courses. These courses fulfill a "rhetoric" requirement for graduates and undergraduates, a "writing intensive" requirement for majors in and outside of English, an elective requirement, or a digital and public humanities requirement. While TPC does in a way overlap with these disciplines and other requirements, the course I teach belongs to an existing plan of study. TPC courses compete with more recognizable forms of English: literature, creative writing, and rhetorical history or theory. The interdisciplinary nature of TPC courses makes it nearly unrecognizable to students and academic advisors who are used to discipline-focused courses. Technical Communication, the course I discuss in this chapter, gets high enrollment when it's offered online, but enrollment becomes harder for an in-person iteration. TPC has tremendous value to students once they dive into these courses. Although many enter my courses with no idea what they will be doing, many end the semester with new perspectives on writing and technology, even if they decide a career in technical communication isn't for them. I describe this context to show one reason TPC can be hidden within colleges and universities. These courses must fit in a nexus of differing ideological values between STEM's need for general education technical writing and the English Department carving out a small space for TPC while maintaining conventional perspectives of what an English major overall does.

Although my work in digital literacies and multimodal composition prepared me to teach technical communication, I wanted to ground the course design in the needs of English majors and STEM majors. I learned from the computer science faculty that their curriculum followed the Accreditation Board for Engineering and Technology (ABET) criteria for the engineering school's accreditation. Among the ABET standards, three areas under Student Outcomes engineering students needed more of were:

- (2) an ability to design solutions for well-defined technical problems and assist with the engineering design of systems, components, or processes appropriate to the discipline;

- (3) an ability to apply written, oral, and graphical communication in well-defined technical and non-technical environments; and an ability to identify and use appropriate technical literature; ...
- (5) an ability to function effectively as a member of a technical team.
  ("Criteria for Accrediting Engineering Programs, 2022–2023," n.d.)

The business school taught conventional writing like memos, resumes, and cover letters. Marketing students learned qualitative data but, ironically, had no creative outlet. What overlapped between the two was that students learned writing to be career ready. I found that what made my role stand out from what other departments already offered was framing technical and professional writing as problem-solving with and from power and privilege that has a significant impact on marginalized people's lives. Multiple theoretical frameworks guided the course design for interdisciplinary teaching, including the rise of engineering justice (Leydens & Lucena, 2018, pp. 19–30), the 3Ps (power-privilege-positionality) framework from technical communication (Jones et al., 2016; Walton et al., 2019), and design justice's emphasis on showing awareness of and collaborating with marginalized communities impacted by technical designs (Costanza-Chock, 2020, p. 6). Taken together, I constructed a course that grounded professional practice in frameworks of social justice. I drew a line from social justice to the civic tech sector as a concrete location to imagine collaborations between industry and social good. Civic technology brings together "the strengths of the private-sector tech world (its people, methods, or actual technology) to public entities with the aim of making government more responsive, efficient, modern, and more just. It also seeks to use digital tech to reimagine interactions among fellow citizens working together, and between those citizens and their governments" (Harrell, 2020, p. 2). Although students would not work with community partners, they would use it as another framework to understand how professional practice overlaps with public concerns of equitable information-sharing for everyday decision-making.

TPC's collaboration with community partners comes with many pitfalls. The hyper-pragmatic approach to service learning (Scott, 2004), even with nonprofits, solidifies the market logics of training in technical communication, or "expectations about professional writers' behavior, performance, and attitudes that are tethered to both macro- and micro-economic concerns about generating capital" (Hashlamon & Teston, 2022, p. 159). A TPC course could teach the dual registers of capitalist and anti-capitalist writing practices, registers that rarely overlap in the corporate world (Hashlamon & Teston, 2022). In addition, service learning can increase the labor of community partners and leave communities unchanged after students have extracted knowledge from those partners, leaving behind tools that aren't useful to the organization.

Considering these institutional and disciplinary contexts, I developed a course design more thoughtful to the kairotic moment of myself, students, and the community organizations as we interact with community partners in Technical Communication as an online asynchronous course. Strategic kairotic interaction with community partners gives more distance from the student and less communication than often desired, yet even this minimal approach has significant gains when coupled with social justice practices. In the following section, I explain the theoretical frame for my course design to expand on these goals.

Technical Communication Pedagogy, Design Thinking, and *Kairos*

Kairos has recently returned to the interests of rhetorical theorists who have debated what it is and how it works (Trapani & Maldonado, 2018). *Kairos* as a concept has often been overlooked among the Greek rhetorical appeals in writing pedagogy. Logos, ethos, and pathos frame writing and argumentation in first-year introductory writing courses instead, although recently *kairos* has been used to frame the use of these three rhetorical appeals. For this chapter, Western classical rhetorical theory, like *kairos*, can be a useful strategy for designing TPC courses and community engagement (Simmons & Grabill, 2007). Scholars like Allen Brizee (2015) and Jessica Edwards (2018) argue for using *kairos* for performing flexible citizenship and deploying critical race theory, respectively. Edwards' pedagogy recalls writing instructors creating and then distributing syllabi and lessons on Twitter (now called X) online to immediately answer egregious violations of constitutional rights and humanity, such as in 2020 during the resurgence of Black Lives Matter amidst George Floyd's murder by police and the rise of anti-Asian hate during the COVID-19 pandemic. Teaching for the moment also summons Pegeen Reichert Powell's kairotic pedagogy. In *Retention and Resistance: Writing Instruction and Students Who Leave*, Powell writes that first-year writing courses follow a chronological pedagogy; this just isn't about creating a linear process but also anticipating what needs to transfer to the next iteration of a writing course in the future. Kairotic pedagogy responds to "the variety of forces in our classroom at any given moment, including the forces in our students' lives that may lead some of them to leave before graduation" (Powell, 2013, p. 14). Powell (2013) thinks about retention in higher education but it's a concern all TPC instructors should share, whether in formal programs or not, and especially for online courses that may add challenges to students. Instructors have an intuitive relationship with time and labor as they design courses; the feature of kairotic pedagogy I add brings to the forefront of our consciousnesses the relationship between time and labor and calls for more careful strategy on when to bring students and community partners together in the interest of learning and retention.

Design thinking is "a combination of a methodology and mindset for innovative problem-solving. It forwards a problem-based approach to innovating solutions by offering guiding principles for choosing and using various methods to understand problems and users" (Tham, 2021, p. 8). Design thinking incorporates five ways of thinking, which can fall into an iterative process: empathize, define, ideate, prototype, and test. Each way of thinking associates with different writing genres and activities, such as user interviews to empathize or mock-ups for ideation. The designing thinking mindset is not meant to be a step-by-step process, but it does fit the logic of writing-tensive courses that follow the writing process by stretching design thinking's five tenets across a semester. I couple design thinking with Sprint Cycles from the Agile project management framework. Project teams of engineers complete a series of goals within a two-week sprint and then report or check-in on their progress. Based on feedback from their clients, engineers create new goals and begin another two-week sprint, and so on.

An eight-week prototype project for community partners suggests rapid assignment submission dates, sometimes occurring multiple times in one week. Short semesters put pressure on instructors to do all they can in little time and do so in ways that busy adult undergraduates may not fall behind. However, design thinking and Agile allow for strategic thinking about how much work to do with regular check-ins with the instructor and the community partner at the end of two weeks. These two design frameworks attempt to respect the labor and time of instructors, community partners, and students. The schedule adapts to adult undergraduates and graduate students who don't have the privilege to focus only on coursework. To be more specific, this kairotic approach to strategic check-ins may respond well to students with disabilities running on crip time (Wood, 2017) and the material challenges that come at the intersections of disability and racial inequality (Bailey & Mobley, 2019). Although leaving an entire week open with no assignments other than a note on progress at the end of the first week may make some instructors uncomfortable, students at the advanced level consistently have been motivated to stay on task. In the following section, I describe the activities that make up a kairotic approach to teaching technical communication in an online asynchronous format.

Design Thinking, Agile, and Kairos in Action

Technical Communication is an introduction to the field broadly defined and focuses on translating students' skills and competencies into communicating technical knowledge for diverse nontechnical audiences. Given my efforts to learn from computer science and business faculty what "technical writing" meant to them, I designed the course to be interdisciplinary. The eight-week project was open to any problem or issue that related to students' interests, goals, and expertise developed from the major. The students could work in small groups or work individually. In this first iteration of the course, taught in spring 2021, the class consisted of mostly English majors, with one English major double-majoring in computer science. In future iterations, this course would attract students in medicine, chemistry, and computer science.

The narrative arch of my course covered three units: theoretical foundations, civic technologies, and the final eight-week project. In unit one, I side-stepped from teaching the links between technical communication and rhetoric as an entryway into the field. Instead, the first unit begins with definitions of technical communication in industry. During the second week, professional technical writers working in health information technology visited the class for a recorded synchronous Zoom meeting to discuss their work, show example writing genres, and explain how they were revising the company's digital service for doctors to have inclusive language. Students asked questions to better learn about the day-to-day lives of these technical writers and about the most important competency for starting a career in the industry. From these interactions, they created a list of principles key for technical communication based on these conversations and then explored how typical professional practices can promote the recognition and liberation of Black technical knowledge (Mckoy et al., 2022, 2020): content design, document design, plain language communication, accessibility, and honoring user's knowledge, desires, and needs. I brought these activities back to their implications for marginalized people, often with real-world scenarios for reflection. In spring 2021, we discussed, for example, adapting communication about the safety of the COVID-19 vaccine to populations who might be resistant to taking vaccines, not just Black people (the legacy of the Tuskegee University Experiment runs across generations of Black families) but also, as one military student explained, veterans because they were ordered to receive vaccines.

Unit Two began the 8-week design challenge with empathy. I juxtaposed industry-based technical communication with civic tech and community engagement to teach the possibilities of the purposes their work could fulfill. For this semester, I developed relationships with three community partners working in neighborhood advocacy with city government, environmental sustainability, and circular economics. As with the industry professionals in unit one, students met with these community organization's representatives to speak about how technical communication worked in their contexts. These partners would be role models or collaborators for the project. Because labor and time were a pressing issue for this online asynchronous course, I gave both students and community partners *the option* to work with one another. Students could connect with a different community partner they knew of or work with someone in their social circle like a family member or friend. Although we discuss racial justice in technical communication, students need to be obligated to do a project related to race and racism. In keeping with Howell's kairotic pedagogy, students could use where they were in life at the present to create a project.

I emphasized both to students and community partners that the result would not be a final version of a product, service, or tool. Instead, community partners would receive a high-fidelity prototype that they could add to their library of project ideas. The student produced the concept, but they could use their own resources and connections to make the concept a reality. On one hand, this approach eased the expectation of students and community partners that they needed something real and it reflected that many project teams in private industry follow the engineering-design-product triad (Harrell, 2020). In this triad, students come from design. We leave the implementation of those tools to the engineers. Only a few students ventured to collaborate with a community partner. Those who did, met with their chosen community partner the following week to learn about their work. Other students conducted research about issues in a community they identified with and/or spoke with a member of that community.

Each mindset in design thinking comes with different writing genres that guide project management. Every two weeks students had a deliverable to submit to myself and to the community partner for optional feedback. This deliverable allowed them to make regular improvements leading up to the tech demo presented to community partners and industry professionals for formal questions and comments. Students who could not present live recorded their presentations and posted them in a shared Google folder for everyone to view. Read Table 13.1 for a complete list of deliverables students created during the eight-week project. Noting that creating deliverables hinges on understanding the user's needs, the students have spent two weeks in Unit Two on qualitative research and analysis of their community or user.

TABLE 13.1 List of assignments for an 8-week design project and descriptions.

1. User counterstory	Drawing from critical race theory, persona stories written as counterstories to reveal the knowledge and assets of users or community partners discovered during research phase
2. Project brief	Formal proposal for idea after analyzing desires, needs, and assets from counterstories and research
3. Prototype with user documentation	High-fidelity version of the solution meeting the needs of the users
4. Usability test report	Reflect on feedback from community partners, peers, and instructor
5. Re-designed prototype	Revised high-fidelity prototype of the solution
6. Tech demo to clients and peers	A slide deck 10–12 slides long presenting the narrative arch of the entire project to community partners, professional writers, and peers for Q&A and feedback

Student Reflections

At the end of the semester, I wanted to learn about students' perceptions of how technical communication and relationality support social justice practices. I invited my students to participate in an IRB-approved one-hour focus group interview in May 2021. Three students agreed to participate in a conversation over a recorded Zoom meeting. These students come from a variety of learning and teaching experiences, and the number of participants in a study cannot make up for the exposure a researcher has *with* the participants. A case study or focus group may garner significant insights that one-hour interviews with fifty participants all at once may not (Small & Calarco, 2022). All three students signed consent forms and gave me permission to use their real names. Here I use only the first names.

- Ashley: a white English graduate teaching assistant. She and a peer collaborated with an organization that provided literacy tutoring for refugee youth.
- Bosten: a white senior English major and Honors student teaching in the Writing Studio. He wrote an employee playbook for his friends' small business.
- Catherine: a Black senior creative writing major. She partnered with a peer to serve a large nonprofit that, among other areas, worked on promoting environmental justice in the metro area.

Based on focus group interviews with these students, I argue that even minimal engagement with community partners promotes powerful lessons about writing in the world for students. The brief, strategically timed interactions still taught students to reconsider the value of writing. They also demanded that academic writing be more engaged with the public. I describe in more detail my analysis below to show the worth of this pedagogical approach to social justice and community engagement.

My technical communication course helped students learn the significant connection between language and audience as a social justice concern. Bosten did not do a project with community partners, but he learned a lot from simply listening to them and practicing social justice practices. Bosten significantly changed his thinking from the start of the semester to the end in reconceptualizing what technical communication was. At first, he expected the course to be more direct: "I saw it very much in this narrow framework of like writing manuals or documents, like how-tos, and related to gadgets or pieces of technology." During his first self-assessment, Bosten had expressed disdain for the first unit's focus on social justice frameworks. In his self-assessment, he wrote about wanting "tools and tips," not more lessons on white supremacy, which he already knew about. He followed up his point arguing that Western knowledge isn't all bad; it has important influences on our lives. However, as he practiced social justice activities such as using plain language communication and identifying other forms of technical knowledge from diverse populations and learning vicariously about his peers' experiences with community partners, Bosten switched his perspective from focusing on writing toward the role of the technical communicators themselves. He explained, "I hadn't really thought about the technical writer being like playing such an important role and being the facilitator of communication between the general public who needs things broken down and the engineers who have to have things written in these complex languages."

Catherine built on Bosten's shaping "complex languages" further by referring to writing with inclusive language practices, especially for Black communities. Her collaborative writing with a community partner on environmental sustainability extended those initial exercises in a high-stake writing project. The organization had multiple focus areas, from aging and health to transportation. Catherine and her peer found focusing on one audience difficult, so they focused on one individual person interested in starting a community garden to address climate change and solar energy. The challenge was producing language: Catherine and her co-writer created an app that included information but that required they "[place] ourselves or [picture] ourselves within the community." Recognizing that "people don't understand how one thing can affect another, like with how gardening can help with their environment. So bringing that knowledge to the table and making an understanding to a community in terms of social justice" was important for Catherine. Community engagement "offer[s] us the opportunity to see it [writing] outside and to be able to look and engage with different audiences. Like each partner may work with a different type of audience."

These connections between language and audience extended to notions of what writing does, a lesson that other students needed to learn. All three students noticed a limitation in how non-English majors and academia valued writing. Often the audience, they argued, was undertheorized and what constituted writing was too narrow. For example, to Bosten English majors can appreciate the nuance of language and writing genres, yet non-majors, in his words, "who aren't necessarily inclined to view in English as a favorable subject get caught up on like trying to check off all the boxes and requirements for an essay and be like, well, I just want to know what do I need to do to get an A? ... What is the right answer for this just like a math equation or a chemistry problem?" However, English majors were still subject to thinking about the professor as the only audience that mattered. Ashley understood this as a graduate teaching assistant. Thinking about creative writing and literature courses, Ashley explained, "I feel like in those classes, you do in a way think about your audience and the different types of people that might be reading your work, but in most English classes, your audience is your teacher. Your audience is the person who's grading your paper." Catherine agreed that "you do write for your professor ... and then you're doing the peer review. So you write it in a more scholarly way, right?" Writing just for the professor, Ashley reasoned, restricted students' "thinking about what the possibilities writing can do for you in your real life." Ashley learned from working with community partners that writing must be purpose-driven and audience-focused and she applied it to her own teaching. As much as her students might dislike learning about the grayness of writing, it was necessary that they "think about your audience, think about how you can use this in future contexts, because that's the most important thing for those who might not be English majors." Although Bosten and Ashley tried to draw a difference between English majors and non-majors, it was clear to me that both kinds of students needed this lesson on expanding audiences and grounding writing in a purpose-driven practice rather than satisfying the norms of academic writing and a professor.

Entering the narratives of community partners, however brief, solidified for students that instructors should blend writing for real-world application and writing in academia. They agreed that writing assignments in academia should be meaningful to students (Eodice et al., 2016) in the following ways: first, combining what they know from previous courses into technical communication and beyond college (Beaufort, 2007; Driscoll & Jin, 2018; Wardle, 2007), and second, having students write for real-world audiences using multimodality. Entering rhetorical ecologies with more diverse writing possibilities means that students need opportunities to justify their rhetorical choices. Ashley explained that being asked to justify their decisions on COVID-19 public service announcements extended rhetorical awareness and reinforced the grayness of content design. In her experience, professors may "discredit" purposeful writing choices when professors should learn about the narrative behind the product. Ashley recalled a student who wrote a research article about women lawyer's casual language in the workplace discredited them among male colleagues. The student used casual language throughout the paper. Ashley reasoned that the student wanted to subvert expectations of the writing genre to argue against sexism in law. Ultimately, the attention to professional practice with community engagement and social justice made these three students think about what writing means and looks like in academics overall and how they can best position themselves as actors in the world. Because writing is purpose-driven or action-oriented, the students themselves have agency in imagining their future careers and the communities they would help with their knowledge.

Design Activities for Kairotic Pedagogy

A kairotic approach to community-engaged teaching can be generative if meaningful assignments and meaningful insights come out of brief interactions with community partners. Instructors may consider the following design exercises in their online courses. Notice that each activity above includes brief yet high-impact points of engagement between students and their community partners.

- Record interviews with one or more community partners; students review the recordings and analyze community partners' needs using empathy maps and rhetorical listening.
- Students share proposal briefs with community partners in a folder in the cloud for feedback.
- Students share low-fidelity prototypes for quick feedback from community partners, the instructors, and classmates.
- Students reveal high-fidelity prototype in a 12-slide pitch deck to community partners.

Conclusion

I briefly demonstrate a kairotic approach to interacting with community partners for an online asynchronous on technical communication and social justice. To honor the labor of both students and community partners who seem to be distant from each other, I strategically consider where students check in with community partners throughout the design thinking methodology. For most of the 8-week project, students receive lots of presence and engagement from me, while community partners read writing genres to give quick feedback and direct the students' moves every two weeks, according to Agile project management frameworks. Technical communication as a problem-focused practice guided by social justice frameworks becomes meaningful and real when students write with and for community partners. Despite minimal engagement throughout the semester, students in my focus group interviews valued their collaborations. As TCP becomes ever more popular and the literature more robust, instructors at smaller schools with a variety of student populations can be more careful in how they implement pedagogies that come down from other well-resourced programs. However, content that binds social justice and community engagement for a short 16 weeks can still give students, not a pathway into technical communication but an expansive view of writing and audience.

References

Agboka, G. Y., & Dorpenyo, I. K. (2022a). Curricular efforts in technical communication after the social justice turn. *Journal of Business and Technical Communication, 36*(1), 38–70. https://doi.org/10.1177/10506519211044195

Agboka, G. Y., & Dorpenyo, I. K. (2022b). The role of technical communicators in confronting injustice—Everywhere. *IEEE Transactions on Professional Communication, 65*(1), 5–10. https://doi.org/10.1109/TPC.2021.3133151

Bailey, M., & Mobley, I. A. (2019). Work in the intersections: A Black feminist disability framework. *Gender & Society, 33*(1), 19–40. https://doi.org/10.1177/0891243218801523

Beaufort, A. (2007). *College writing and beyond: A new framework for university writing instruction.* University Press of Colorado.

Bourelle, T. (2014). Adapting service-learning into the online technical communication classroom: A framework and model. *Technical Communication Quarterly, 23*(4), 247–264. https://doi.org/10.1080/10572252.2014.941782

Brizee, A. (2015). Using Isocrates to teach technical communication and civic engagement. *Journal of Technical Writing and Communication, 45*(2), 134–165. https://doi.org/10.1177/0047281615569481

Cleary, Y. (2021). Fostering communities of inquiry and connectivism in online technical communication programs and courses. *Journal of Technical Writing and Communication, 51*(1), 11–30. https://doi.org/10.1177/0047281620977138

Costanza-Chock, S. (2020). *Design justice: Community-led practices to build the worlds we need.* MIT Press.

Criteria for Accrediting Engineering Programs, 2022–2023. (n.d.). ABET. Retrieved December 23, 2023, from https://www.abet.org/accreditation/accreditation-criteria/criteria-for-accrediting-engineering-programs-2022-2023/

Driscoll, D., & Jin, D. (2018). The box under the bed: How learner epistemologies shape writing transfer. *Across the Disciplines, 15*(4), 19–20. https://doi.org/10.37514/ATD-J.2018.15.4.19

Dumlao, R. J. (2023). Collaborating successfully with community partners and clients in online service-learning classes. *Journal of Technical Writing and Communication, 53*(3), 218–239. https://doi.org/10.1177/00472816221088349

Edwards, J. (2018). Race and the workplace: Toward a critically conscious pedagogy. In A. M. Haas & M. F. Eble (Eds.), *Key theoretical frameworks: Teaching technical communication in the twenty-first century* (pp. 268–286). Utah State University Press. https://doi.org/10.7330/9781607327585.c011

Eodice, M., Geller, A. E., & Lerner, N. (2016). *The meaningful writing project: Learning, teaching and writing in higher education.* Utah State University Press.

Francis, A. M. (2018). Community-engaged learning in online technical communication classes: A tool for student success. In G. Y. Agboka, & N. Matveeva (Eds.), *Citizenship and advocacy in technical communication: Scholarly and pedagogical perspectives* (pp. 223–242). Routledge.

Garrett, R., Simunich, B., Legon, R., & Fredericksen, E. E. (2023). *CHLOE 8 student demand moves higher ed toward a multi-modal future, the changing landscape of online education, 2023.* Quality Matters and Encoura Eduventures Research.

Grabill, J. T. (2004). Technical writing, service learning, and a rearticulation of research, teaching, and service. In T. Bridgeford, K. S. Kitalong, & D. Selfe (Eds.), *Innovative approaches to teaching technical communication* (1st ed., pp. 81–92). Utah State University Press.

Harrell, C. (2020). *A civic technologist's practice guide.* Five Seven Five Books.

Hashlamon, Y., & Teston, C. (2022). Teaching participative justice in professional writing. *Technical Communication Quarterly, 31*(2), 159–174. https://doi.org/10.1080/10572252.2021.2000031

Hewett, B. L., & Bourelle, T. (2017). Online teaching and learning in technical communication: Continuing the conversation. *Technical Communication Quarterly, 26*(3), 217–222. https://doi.org/10.1080/10572252.2017.1339531

Johnson-Eilola, J., & Selber, S. A. (2013). *Solving problems in technical communication.* The University of Chicago Press.

Jones, N. N. (2016). The technical communicator as advocate: Integrating a social justice approach in technical communication. *Journal of Technical Writing and Communication, 46*(3), 342–361. https://doi.org/10.1177/0047281616639472

Jones, N. N., Moore, K. R., & Walton, R. (2016). Disrupting the past to disrupt the future: An antenarrative of technical communication. *Technical Communication Quarterly, 25*(4), 211–229. https://doi.org/10.1080/10572252.2016.1224655

Leydens, J. A., & Lucena, J. C. (2018). *Engineering justice: Transforming engineering education and practice.* John Wiley & Sons; IEEE Press.

Mckoy, T., Shelton, C. D., Sackey, D. J., Jones, N. N., Haywood, C., Wourman, J., & Harper, K. C. (2020, October 5). CCCC Black Technical and Professional Communication Position Statement with Resource Guide. *Conference on College Composition and Communication.* https://cccc.ncte.org/cccc/black-technical-professional-communication/

Mckoy, T., Shelton, C. D., Sackey, D. J., Jones, N. N., Haywood, C., Wourman, J., & Harper, K. C. (2022). Introduction to special issue: Black technical and professional communication. *Technical Communication Quarterly*, *31*(3), 221–228. https://doi.org/10.1080/10572252.2022.2077455

New London Group (1996). A pedagogy of multiliteracies: Designing social futures. *Harvard Education Review*, *66*(1), 60–92.

Nielsen, D. (2016). Facilitating service learning in the online technical communication classroom. *Journal of Technical Writing and Communication*, *46*(2), 236–256. https://doi.org/10.1177/0047281616633600

Powell, P. R. (2013). *Retention and resistance: Writing instruction and students who leave*. Utah State University Press. https://search.ebscohost.com/login.aspx?direct=true&AuthType=sso&db=nlebk&AN=696625&scope=site&custid=078-820

Scott, J. B. (2004). Rearticulating civic engagement through cultural studies and service-learning. *Technical Communication Quarterly*, *13*(3), 289–306. https://doi.org/10.1207/s15427625tcq1303_4

Simmons, W. M., & Grabill, J. T. (2007). Toward a civic rhetoric for technologically and scientifically complex places: Invention, performance, and participation. *College Composition and Communication*, *58*(3), 419–448.

Small, M. L., & Calarco, J. M. (2022). *Qualitative literacy: A guide to evaluating ethnographic and interview research*. University of California Press.

Tham, J. C. K. (2021). *Design thinking in technical communication: Solving problems through making and collaboration*. Routledge.

Trapani, W. C., & Maldonado, C. A. (2018). *Kairos*: On the limits to our (rhetorical) situation. *Rhetoric Society Quarterly*, *48*(3), 278–286. https://doi.org/10.1080/02773945.2018.1454211

Walton, R., & Agboka, G. Y. (2021). *Equipping technical communicators for social justice work: Theories, methodologies, and pedagogies*. University Press of Colorado.

Walton, R., Jones, N. N., & Moore, K. R. (2019). *Technical communication after the social justice turn: Building coalitions for action*. Routledge.

Wardle, E. (2007). Understanding "transfer" from FYC: Preliminary results of a longitudinal study. *WPA: Writing Program Administration*, *31*(1/2), 124–149.

Wood, T. (2017). Cripping time in the college composition classroom. *College Composition and Communication*, *69*(2), 260–286.

Zamparutti, L. (2022). The "antenarrative" in online-asynchronous technical communication courses: A social justice approach to teaching. *2022 IEEE International Professional Communication Conference (ProComm)*, 392–396. https://doi.org/10.1109/ProComm53155.2022.00079

14

DESIGNING WITH CARE

A Cultural Rhetorics Praxis of Care for Digital Storytelling Projects About Reproductive Justice

Danielle Marie Koepke

Recent attacks on abortion laws and policies have drawn national attention to the fight for reproductive rights. However, Black women, Indigenous women, Women of Color, and LGBTQ+ people have long been fighting for reproductive justice in the USA and around the globe. Reproductive justice is a human rights framework under the umbrella of social justice. It is not a single-issue women's rights movement ("Reproductive Justice," n.d.); it makes visible the intersections between reproductive issues and other issues such as immigration rights, fair wages, housing, quality education, and safe neighborhoods (Ross et al., 2017). While it is a wide-ranging movement, many efforts have focused on abortion since June of 2022, when the *Dobbs v. Jackson Women's Health Organization* case was brought before the U.S. Supreme Court and the court decided that the right to abortion was not held up by the constitution. This ruling overturned *Roe v. Wade*, which since 1973 had protected a woman's choice to have an abortion (The New York Times, 2022). This court ruling allowed states to make their own decisions and regulations regarding abortion care and services. In the initial months that followed the ruling, 13 states completely banned abortions while others created more harsh gestational limits. Some questioned life-saving medical procedures and others banned the use of medicines to end pregnancy such as mifepristone and misoprostol and/or access to levonorgestrel emergency contraception, commonly known as Plan B or the morning after pill. In states where abortions were still legal, clinics became overwhelmed. In my state, Wisconsin, an 1849 criminal abortion law was put back into effect for over a year. One recent study estimated that from January to June of 2023,

DOI: 10.4324/9781003469995-18

1,503 people had to carry out unintended pregnancies in Wisconsin alone (Green & Higgins, 2023). These hardships have fallen disproportionately on Black people and People of Color (McGinn Valley et al., 2023), at increased levels. The reproductive justice landscape is laden with unmet healthcare needs as it continues to shift and change in Wisconsin and across the nation.

In this political and sociotechnological terrain, digital storytelling has expanded exponentially as an advocacy tool for social justice movements. There are many affordances of using a story as a method to call people to action. Stories allow the audience to connect their own personal experiences to the storyteller's. Stories allow the audience to see and feel the impacts of seemingly abstract policies. Stories are relatable. Sharing diverse stories of reproductive [in]justice expands notions of who needs and deserves reproductive, sexual, and gender freedoms. However, scholars and community partners should be wary of the high rhetorical velocity (Ridolfo & DeVoss, 2009) of these often-vulnerable stories. Such a rapid circulation disallows control of how a story is received and how it might be used, changed, and shared with alternative purposes than originally imagined. Scholars of technical and professional communication (TPC) are uniquely positioned to design, support, and guide community storytelling projects and can draw on their expertise in justice-oriented technologies to guide advocacy projects as they are circulated in digital public spaces.

In this chapter, I reflect on the efforts of designing a digital storytelling project about the stories of the Promotores de Salud, Latinx health promoters working for reproductive justice in their own communities across Wisconsin. *Cuentos de Confianza: A Community Writing Project for Reproductive Justice* is a bilingual digital storytelling project that highlights stories written by six Latina promotoras about their lived experiences of reproductive injustices and of advocating for sexual and reproductive education and health. I worked in close collaboration with Dr. Rachel Bloom-Pojar (UW-Milwaukee), the promotores, and undergraduate translators to design the site, curate the content that framed the stories, and circulate the stories for the purpose of the promotores, which was to celebrate the important work they do and to educate their friends and families on topics of reproductive justice. Throughout, I grappled with how to care for their stories, not as flattened words on a screen but as vulnerable, embodied experiences. To center their needs and goals, I adopted practices from across disciplines that supported flexible, vulnerable, slow-moving-but-active care. By constellating reproductive justice, cultural rhetorics, and design justice, I built a cultural rhetorics praxis of care that supports community-engaged digital storytelling projects about reproductive justice. This chapter examines how such praxis of care can offer practical actions for caring for vulnerable stories and storytellers that align with social justice initiatives in TPC. I end by calling scholars in TPC to critically reflect on how they can move towards valuing story and relationships more than research and design methods, towards centering community knowledge and experiences, and towards pursuing new and alternative actions to best care for storytellers. Ultimately, I hope to demonstrate how scholars in TPC can actively support and deeply care for community-engaged projects that advocate for social justice initiatives.

Cuentos de Confianza

For the past four years, I have collaborated with a group of the promotores, Dr. Rachel Bloom-Pojar (UW-Milwaukee), and undergraduate student translators to create a bilingual digital storytelling project that highlights the experiences of the promotores and situates their efforts within the reproductive justice movement. *Cuentos de Confianza: A Community Writing Project for Reproductive Justice* originally hosted six stories written and recorded by promotoras (female-identifying health promoters) from across Wisconsin. While the project now has 13 stories, and continues to transform, in this section I offer some background on the promotores and the original iteration of the project.

The Promotores de Salud program in Wisconsin, which has been actively growing for over 20 years, seeks to provide culturally responsive sex education to Latinx communities across the state. The largest groups of promotores are in Milwaukee and Madison, but they have also been expanding outreach in smaller cities and more rural areas. The promotores work as paid consultants through the community education department of Planned Parenthood of Wisconsin (PPWI) to "provide culturally competent sexuality education in Spanish to their own social networks" at fiestas caseras (home health parties) with cross-generational audiences ("Promotores de Salud and Health Promoter Programs," n.d.). Home health parties are hosted by a community member who invites family and friends to come for food, fellowship, and educational conversations about sexual and reproductive health and wellness topics. The promotores also support their communities in navigating immigration and citizenship issues, such as providing voting registration information. During the COVID-19 vaccine rollout, they helped with educating the community, correcting misinformation, and promoting mobile vaccination clinics. When someone moves in, they give them a tour and connect them with resources. The promotores are essential to the reproductive justice movement (NLIRH, 2010, p. 8) because they work within their own communities, in which they've spent time building *confianza*. Confianza translates to "trust" or "confidence" in English, but more dynamic meanings are lost in translation. Entering into confianza is the foundation of everything the promotores do. While this practice is complex and fluctuates over time and context, it tends to incorporate building relationships, sharing stories, and being present in the moment (Bloom-Pojar & Barker, 2020). This slow-grown, relational trust allows them to "bridge a gap" between the U.S. healthcare system and the needs of immigrant families (Bloom-Pojar & Barker, 2020, p. 15), which often include challenges such as transportation to services, understanding the healthcare infrastructure, and translating medical jargon into everyday language.

Through the confianza they foster, they are seen as trusted people with whom community members can share traumatic or vulnerable experiences such as infertility, intimate partner violence, or child abuse. The promotores take on this heavy emotional labor, which is far beyond what they are paid for because they feel a community responsibility to fill these gaps in care left by healthcare institutions (Bloom-Pojar & Barker, 2020, p. 92). While these labors are heavy, the promotores see them as essential to empowering their communities on the path towards reproductive justice ("Cuento de Elida;" "Cuento de Angeles"). María Barker, director of the promotores, reflected on how the promotores, many of whom are women, needed to learn to accept that caring for themselves first allowed them to better care for others: "At first, our health promoters could not understand the reproductive justice movement because they had always put others first, and they always came second, third, or fourth. The concept of needing to take care of themselves first was not anything they had considered, but when told how important it is for them to be well so they can take care of others, it made sense to them" (Bloom-Pojar & Barker, 2020, pp. 89–90). Knowing this, Dr. Bloom-Pojar and I sought to care for *them* and *their* experiences as we embarked on this project together. We felt accountability to create a project that allowed the opportunity for them to care for *themselves*.

The stories for *Cuentos* were developed through a voluntary writing class taught by Dr. Bloom-Pojar and assisted by me in 2021–2022. We designed the course to offer time and space for the writers to reflect on their layered identities related to their reproductive justice work and to their own embodied experiences of reproductive [in]justice. Publishing their stories on the site was optional, and we checked in many times about what, if anything, the writers wanted to publish. We had conversations about the potential harms that could come from sharing personal stories publicly and digitally, about whether to use full names, about including photos of themselves and/or of their families, and about other privacy and security concerns. Ultimately, it was up to each writer to decide what they wanted to share, and how, knowing that they could change or withdraw their story at any time if they changed their mind. After we hosted a community launch event for close friends and family to celebrate the writer's stories, *Cuentos* was implemented into some of the home health parties as a supplemental educational tool.

The writers had a range of feelings as they shared their stories in person with groups from their communities to open conversation on difficult topics surrounding sexual and reproductive health. Until you put your own vulnerable story out into the world, it is hard to imagine just how it feels to read back over those raw experiences. You relive them. Because I wanted to understand this feeling and because I wanted to reciprocate in some small way, I opened my dissertation about *Cuentos* with my own vulnerable story. As I wrote, edited, and relived the experiences of getting pregnant at 17 and dealing with the manipulative, aggressive, controlling biological "father," I had all these complicated thoughts about putting my own story out there. I wanted to be taken seriously and I also wanted people who read my dissertation, even if it was only ever my committee, to understand me as more than just an academic. I wanted readers to know my own embodied connections to reproductive justice, but I didn't want to center my own experience as a cis white able-bodied and educated woman. I worried about backlash because I was in the middle of a court case regarding that child, but I didn't think I could rightfully write about the stories of others without sharing my own. Making a choice to share a story is laden with complicated and competing emotions, even if it feels like it is supporting a movement. Working so closely with vulnerable stories that I read and listened to repeatedly caused me to slow down and approach all decisions with deep care for the storytellers.

Scholarship that Supported My Actions

Throughout this unique community-engaged project, I leaned on scholarship across disciplines as I sought to care for these storytellers and their stories. In this section, I'll share some of the theoretical underpinnings that supported me. However, I caution readers against thinking these three areas—reproductive justice, cultural rhetorics, and digital rhetorics—are isolated. They are interrelated and crossover with one another, and they themselves rely on other areas of scholarship, especially Indigenous, feminist, and queer theories.

Reproductive Justice

Reproductive justice is a term that was coined by a group of Black women who were tired of being excluded from a racist, classist [white] women's movement in the 80s and 90s ("Reproductive Justice," n.d.). These women, who eventually formed the national reproductive justice organization named SisterSong, sought policies that recognized and supported inter-sectional and marginalized identities (Crenshaw, 1989). Today, there are four main tenets to reproductive justice: the right to not have children, the right to have children, the right to parent children in safe and healthy environments, and the right to sexual and gender autonomy. Reproductive justice is a demand for universal human rights for all people, especially the multiply marginalized. While national attention is needed for large-scale changes to happen, reproductive justice initiatives are most success-ful when the everyday work is centered on local community concerns (see Cook, in Silliman et al., 2004, for one example). As such, the *Cuentos* pro-ject necessarily required adherence to reproductive justice principles and to community-engaged practices for ethical collaboration. Reproductive justice scholarship has been documenting the transformative action hap-pening in communities across the nation (Ross et al., 2017; Silliman et al., 2004). Some scholars in rhetoric and writing studies and the rhetorics of health and medicine have also been advancing issues of reproductive jus-tice and the ways in which scholars can support and practice activism to varying degrees (Grobman & Mutnick, 2020; Novotny & De Hertogh, 2022). Marginalized storytellers are often asked to share their vulnerable stories for social justice endeavors, whether academic or not, but it can be hard to convey (or imagine) all the potential risks to security, privacy, and well-being after a story is out in public. Reproductive justice activism and scholarship taught me how to center the embodied experiences of the marginalized community I was working with and how to prioritize action over theory-making.

Cultural Rhetorics

Cultural rhetorics is my orientation to rhetorical study. I see all commu-nication and meaning-making as deeply steeped in culture, and vice versa (Powell et al., 2014). We cannot pluck an artifact out of its cultural con-text and expect to analyze it or understand it. This is especially true of stories. First, as a cultural rhetorician, I view stories as important theories about the world and how to be in it. The story as theory orients me to lis-ten to stories and honor the embodied experiences they describe (Arellano et al., 2021, p. 5). Second, being aware of relations is essential to research, teaching, and community work because we literally are the sum of our relations (Wilson, 2008). Recognizing my relations to other people, the land, my body, and reproductive justice work calls me to actionable reci-procity and accountability to the communities I am in relation with. It also makes visible my own embodied relations to reproductive justice (Novo-tny & Opel, 2019). Third, I engage in constellative work, drawing lines of connection among the expertise and knowledges of the promotores, of my collaborators, and of myself and making way for coalitional work. I was also able to draw on the perspectives of queer and feminist traditions as well as scholarship on care that was outside of rhetoric and writing studies to explain the practices I was enacting. This allowed me a "horizontal freedom to move across boundaries [...] and draw on multiple theories across fields, as well as my own relation to the issue" (Arellano et al., 2021, p. 17). Finally, decolonial practices are necessary if we are to move towards real, material justice. Engaging in decolonial practices as academics requires us to recognize systems of power and to actively participate in reparations and other actions towards justice (Tuhiwai Smith, 2012). We strove to break down power dynamics, always allowing the promotores to lead the way and striving to avoid interference or oversight from our university or from PPWI on this project. I draw on these principles differently at different times, but relations and stories have been especially prevalent throughout the life of the Cuentos project. Cultural rhetorics isn't quick and easy (Arellano et al., 2021, p. 19). It takes time to build relationships, listen to stories, and foster a sense of trust and it will not look, sound, or feel the same to everyone (Riley Mukavetz, 2014).

Digital Rhetorics and Design Justice

Working on a digital project necessitates rhetorical interrogation of and engagement with the impacts of design and digital technologies on sharing stories. There has been growing momentum towards action-oriented social justice approaches in the field of TPC (Haas & Eble, 2018; Walton et al., 2019). With it, myths of communication technologies being neutral have been shattered (Benjamin, 2019; Jones, 2016). Digital rhetoric is "the digital negotiation of information – and its historical, social, economic, and political contexts and influences – to affect change" (Haas, 2018, p. 412), and making rhetoric plural merely accounts for the multiple rhetorical traditions that influence these digital negotiations. As society and technology become more and more tightly bound, it is essential to interrogate how digital spaces are designed and what methods of communication happen or are encouraged based on the design. My own digital design process has leaned heavily on design justice, which "rethinks design processes, centers people who are normally marginalized by design, and uses collaborative, creative practices to address the deepest challenges our communities face" ("Design Justice Network Principles," 2018). When working on digital community projects, we must be wary of "which humans are prioritized in the [design thinking] process" (Benjamin, 2019, p. 176). Design justice seeks to name and dismantle the unequal power dynamics at work in the design of technologies and recognizes design as "a way of thinking, learning, and engaging with the world" (Costanza-Chock, 2020, p. 15). Digital design should be utilized to empower communities, recognizing expertise based on lived experiences, not simply academic theories. While it may often be unintentional, designers tend to only take a small group of people into account when creating technologies. For example, it is now widely recognized that facial recognition is based on and works best for people who are white/Caucasian and male (Costanza-Chock, 2020). This leads to exclusionary technologies that harm and oppress most groups of people, especially exerting a negative impact on marginalized communities. Practicing design justice has meant not only including the community writers in the design of Cuentos but also creating a sustainable tool that they can be proud of and use for their own desired outcomes. These three areas of scholarship guided my actions as I tried, failed, and learned how best to care for the stories and storytellers of *Cuentos*.

A Praxis of Care for Digital Storytelling Projects about Reproductive Justice

As I kept coming back to the question, "how best to care for these stories?" I dove into scholarship that defined, unpacked, or critiqued "care." In her book, *Living a Feminist Life*, Ahmed (2017) etymologically breaks down the word "care" into the Old English connotations of "sorrow, anxiety, [and] grief" (p. 169). These words suggest that when someone is full of care, they are anxious about potential ills, breakages, or failures. Ahmed shows how breakage is not the end of a thing but opens possibilities for new things. For Ahmed, care is an acceptance of a fragile, vulnerable co-existence with other fragile individuals by those who deviate from heteronormativity. Piepzna-Samarasinha (2018) describes care as networked webs that form among disabled folks in which they "[contribute] as they can, not necessarily 'equally' or always" (p. 28). These webs of care are interconnected and, when they hold together, they serve a meaningful purpose. But they are also delicate; they do not last forever. Feminist rhetoricians Novotny and Opel (2019) reclaim care from mere feminist critique to meaningful action. Care, they claim, is a responsive and reflective action (p. 99). It is an activist response to a community need. Puig de la Bellacasa (2017) speculates on an action-oriented and multifaceted approach to care: care as labor, work, or activity, care as an affect or feeling, and care as a political ethic (p. 204). Care is not just a warm fuzzy feeling (Ahmed, 2017); it is maintenance work for collective survival that is "manifested in practice – action, labor, work – it is integral to our ways of doing"

(Malatino, 2020, p. 41). Care is a networked web of doings, affects, and ethics towards individual and group continuity.

Care is messy. Labors of care are not necessarily mutually reciprocated, and often they are unevenly distributed. The U.S. has a history of using care to control or erase disabled people (Piepzna-Samarasinha, 2018). Even family members may use care as a means of control, whether ill or well-intentioned (Kelly, 2013). Similarly, the rhetoric of care can be used to justify violence against marginalized communities who aren't living up to heteronormative standards in the eyes of those in power (Hong, 2015). Historically, care work in the USA has been done mostly by third-world women who have been paid and valued little even though they have done important work of nurturing future generations (Katz, 2022). Care gets frivolized and monetized, too, like when self-care is mobilized as merchandize or as an ad tactic. Care is fraught with conflicting connotations. As such, care ethics must include situational awareness (Perhaps, a rhetorical awareness). We must grapple with *how to care* within the imperfect, changing, fragile webs we find ourselves entangled in.

Drawing from these knowledge, and the areas of scholarship above, I developed a cultural rhetorics praxis of care for digital storytelling projects about reproductive justice that centered on vulnerability, slow and flexible actions, and community. A cultural rhetorics praxis of care values the *doings, feelings, and ethics* of care by putting the principles of cultural rhetorics into reflective action. First, valuing story as theory highlights the lived experiences of community collaborators and calls scholars to act on the implications of circulating their stories in digital spaces, even if with good intentions. This practice holds researchers responsible for caring about stories as extensions of storytellers' embodied experiences. Second, honoring relations brings attention to our intertwined embodied realities and calls us to respond with accountability and reciprocity. Scholars should "do the work needed in the communities that surround us, whether that work includes academic research outcomes or not" (Grant & Walker, 2023, p. 11). This practice also gently pushes researchers to continually pass ownership of projects back to the community. Third, building constellations of knowledge and expertise supports community ways of knowing and fosters genuine collaboration. Constellation is a space-making practice of care that creates the necessary room for consideration of the researcher's embodied experiences in relation to the research project and in relation to the community while also valuing the community's knowledge, practices, and ways of being. Fourth, decolonial engagement decenters academia and recenters the community. Refusing certain research methods and academic timelines is imperative to live out this actionable care. It also means accepting "no's" and to be sensitive to times when not saying "no" might not necessarily mean "yes" (Itchuaqi-yag, 2021, p. 46). This praxis supports social justice initiatives involving community-engaged story projects and supports scholars who may feel their research, deliverables, and community work doesn't resemble traditional research methods and methodologies (Amidon et al., 2023).

Designing with Care

As I designed the *Cuentos* site, care was at the forefront of my mind. Because this was a community-engaged project, the most important way to demonstrate care was to involve the writers in as much of the design and curation process as they wanted. When researching platforms that offered website creation and design, I chose to use Wix because of its capabilities and usability. Wix also allowed me to do a lot on the free version before we had to commit to a paid subscription and had a good mobile site editor. The digital storytelling site has been through multiple iterations, each of which I designed from scratch. While doing this work has not involved hard coding, it has involved tedious layout changes and updates; creation and movement of page items like container boxes, text, and decorative elements; and lots of trial and error on everything from color schemes to font type to formatting to layout of the individual pages. Because of the bilingual audience, unique purpose, and planned circulation of the site, a template just wouldn't work for *Cuentos*. Most templates come with goals like selling merchandise, promoting a business, or displaying art exhibits. Our purpose didn't fit any of the predesigned templates or categories. We needed to be able to customize everything across the pages, so that's what I did. After conversations with Dr. Bloom-Pojar and María Barker, I presented a design to them as a starting point so we could find out what we collectively wanted and what we didn't want. After receiving feedback on what the overall feel of the site should be along with what content might be helpful for visitors to the site, I started over. We didn't quite yet know how exactly the digital site would be implemented into community use as this was before the community writing class took place. The second iteration had many more page ideas like a page to explain the fiestas caseras, a page for civic engagement, and even a pedagogical statement for the community writing class. Neither of these earlier stage designs prioritized Spanish because I was still researching how to best do that. The third iteration is our current site, though it has been through many tweaks and updates. Below, I discuss four literal actions of designing Cuentos through a cultural rhetoric praxis of care. They do not function as 1:1 with the four principles but instead are each impacted by multiple principles of the praxis (Figure 14.1).

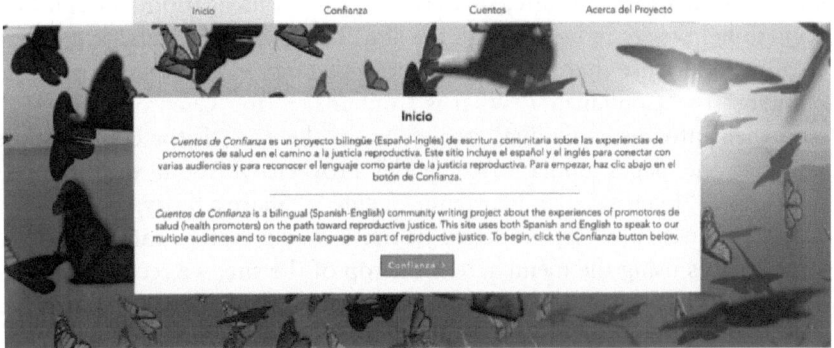

FIGURE 14.1   A screenshot of the *Cuentos* landing page, "Inicio."

First, throughout the design process, I did my best to center Spanish-speaking users in design because the target audience was the promotores' own Latinx communities and because language access intersects with reproductive justice. I built the Spanish and English elements at the same time to avoid force-fitting something from English to Spanish. While this led to more time-consuming work, it has allowed me to create a deliverable that the promotores feel is theirs to have for themselves, to share with their target audience, and to use at their home health parties. "Inicio," which in English translates to start or beginning, is the site's landing page. This is the page you will find yourself on when you open the website. This page is in both Spanish and English, with Spanish purposefully on top. Initially, I tried to have Spanish and English together on every page. But I found that it was confusing and overwhelming to read. I also worried about the length of these pages on the mobile version of the site. I separated the Spanish and the English, always privileging Spanish. I created tabs at the top of the site in Spanish, accompanied by English subtabs that appear when you hover over each tab. Each of the writers wrote their story in Spanish first, so each of the English story pages was built from the Spanish version, both in content and in design. Our translator worked with each of the writers to translate their Spanish cuento into English and to make sure that it reflected their personal voice in both languages through multiple meetings. Each story is also accompanied by an audio recording of the writer reading their story in Spanish on both the Spanish and English pages. We encourage English speakers to listen to the cuento (story) in Spanish as they follow along with the English text.

Second, I created navigation that was simple and easy to follow so the design wouldn't inhibit user experience. One issue this created was a multitude of pages to navigate, so I created guided navigation through buttons to help users move through the site. The goal was to create a digital exhibit experience that users would move through in a certain order: from "Inicio," to "Confianza"/"What is Confianza?" to "Cuentos"/"Stories," and then into the individual stories. From there, a visitor could choose to view the pages under "Acerca del Proyecto"/"About the Project": "Eventos"/"Events," "Comparte un Cuento"/"Share a Story," and "Justicia Reproductiva"/"Reproductive Justice." While visitors can navigate to other places using the menu across the top of the site, we recommend following our suggested navigation so that visitors read about confianza and our care statement before engaging with the stories. Once on a page, you can also click on the pink button in the upper right-hand side of the screen to toggle between the Spanish and English versions. In the image below, you would click on the "Español" button to go to the Spanish version of the page. We especially wanted to be sure that visitors to the site read the care statement on the "Confianza" page, which was developed in the community writing class by the writers with input from Rachel and me. We've also added a resources tab with vetted sources that offer bilingual information and support on topics related to the stories.

Third, I framed the stories within a collaborative care statement and background information to couch them in a few layers of protection. While someone could choose not to follow the site's navigational structure, my hope is that most readers will be prepped so that by the time they are engaging with the actual stories, they will understand their context and will be prepared to care for them. After reading the care statement, visitors are directed to click on the pink *Cuentos* button which leads to the "Cuentos" or "Stories" page hosting the writer bios, images, and access to their stories. Writer bios and accompanying images are in alphabetical order. After showing them an example, each writer put together a bio and a picture of their choosing. The writers chose how they wanted to represent themselves in their bio. As we decided which pages helped to frame and contextualize the stories, we decided to move "Reproductive Justice" to a nested page under "Acerca del Proyecto"/ "About the Project." We didn't want it to dissuade visitors from the site or overwhelm them with information. Rachel and I concurred that while a secondary audience such as scholars might be interested in it as a framing page, it didn't seem central to how we envisioned the target audience engaging with the project.

Fourth, I did everything I could to deconstruct power dynamics and give control of the design, purpose, and circulation of the project to the promotores. I communicated with my collaborators so that especially the writers were informed about their consent to share their personal stories on Cuentos, how they chose to do that, and future avenues for circulation. Dr. Bloom-Pojar and I pulled together what we thought should be in a justice-oriented consent form. In Spanish, the terms were explained to the promotores, but there were also multiple conversations about the implications of sharing personal stories and personal identifying information in a digital public space. We made sure to include multiple check-ins, during which times I showed the site, they reviewed their bios and stories, and were able to make and/or suggest changes as we moved into a slow rollout of the site. The promotores were excited to be able to share their stories; at the same time, I was anxious to avoid any kind of harm that might come through public access to the site. One of our roles as scholars who bridge the university and the community is gatekeeper so that access does not lead to exploitation (Itchuaqiyag, 2021). We held meetings to understand how UW-Milwaukee and PPWWI viewed (or sought to lay claim to) the project and researched how to keep the project within the community in the future. Taking on these labors allowed the promotores to focus on integrating the project into their communities and gave them peace of mind about the longevity of ownership.

One Design Doesn't Fit All

It is difficult to offer a template of design or a specific learning exercise to emulate this design work in other community-engaged projects for social justice initiatives. Cultural rhetorics and community-engaged scholarship both encourage scholars to "resist the notion that community-based research should be replicable. Relationships are not replicable" (Riley Mukavetz, 2014, p. 121). We must take actions based on the relationships, cultural context, and the needs of the community. For *Cuentos*, we sought to build relationships and support community needs and goals. From there, the community class organically came to be. From there, the digital storytelling project developed. We didn't force the promotores to write certain stories so that they could be used in the home health parties. We held everything loosely as we sought to foster a collective vision of the project, and we continue to do so. For teachers and researchers within TPC, what I offer in parting are a few questions to germinate critical reflection about how to practice care as you work on vulnerable community projects and possibly share them in the classroom:

1 In what ways is the digital design and/or technology reflecting the priorities of the community?
2 What actions are you taking to protect, support, or center community storytelling projects?

3 Which practices of care lead your circulation of the project in digital spaces?
4 How are you integrating the project into the classroom while attending to care for students, stories, and storytellers?

Caring for the stories and storytellers of social justice projects requires situational awareness and flexible, vulnerable, and slow-moving actions. This is not easy work, and it gets harder when stories are quickly and widely circulated in digital spaces. These questions can provide researchers with a creative and critical reflection process to think through the *how* of implementing a cultural rhetorics praxis of care specifically on digital storytelling projects with community collaborators on vulnerable topics such as reproductive justice.

References

Ahmed, S. (2017). *Living a feminist life.* Duke University Press.

Amidon, T. R., Moore, K. R., & Simmons, M. (2023). Community engaged researchers and designers: How we work and what we need. *Communication Design Quarterly, 11*(2), 5–9. https://dl.acm.org/doi/10.1145/3592356.3592357

Arellano, S. C., Bretnell, L., Hsu, J., & McGee, A. (2021). Ethically working within communities: Cultural rhetorical methodologies principles, *Constellations: A Cultural Rhetorics Publishing Space, 4,* 1–25, https://constell8cr.com/conversations/cultural-rhetorics-methodologies/

Benjamin, R. (2019). *Race after technology: Abolitionist tools for the new Jim code.* Polity.

Bloom-Pojar, R., & Barker, M. (2020). The role of confianza in community-engaged work for reproductive justice. *Reflections: A Journal of Community-Engaged Writing and Rhetoric, 20*(2), 84–101. https://reflectionsjournal.net/wp-content/uploads/2020/12/V20.N2.BloomPojarBarker.pdf

Costanza-Chock, S. (2020). *Design justice: Community-led practices to build the worlds we need.* MIT Press.

Crenshaw, K. (1989). Demarginalizing the intersection of race and sex: A Black feminist critique of antidiscrimination doctrine, feminist theory and antiracist politics. *University of Chicago Legal Forum, 1989*(1), 139–167. https://chicagounbound.uchicago.edu/cgi/viewcontent.cgi?article=1052&context=uclf

Cuentos de Confianza: A Community Writing Project for Reproductive Justice. n.d. https://www.cuentosdeconfianza.com/

Design Justice Network (2018). *Design justice network principles. Allied Media Projects,* https://designjustice.org/read-the-principles

Grant, C., & Walker, D. (2023). Designing public identity: Finding voice in coalitional technical writing with Black-led organizations. *Communication Design Quarterly, 11*(2), 10–17. https://dl.acm.org/doi/10.1145/3592356.3592358

Green, T., & Higgins, J. (2023, December 14). UW professors: Dobbs forced at least 1,500 unintended births, causing harm to Wisconsin communities. *Wisconsin State Journal.* https://madison.com/opinion/column/uw-professors-dobbs-forced-at-least-1-500-unintended-births-causing-harm-to-wisconsin-communities/article_5ca609da-9912-11ee-8970-43a390d8a11c.html

Grobman, L., & Mutnick, D. (2020, Fall/Winter). *Reflections special issue. Reflections: A Journal of Community-Engaged Writing and Rhetoric, 20*(2), 1–254.

Haas, A. (2018). Toward a digital cultural rhetoric. In J. Alexander, & J. Rhodes (Eds.), *The Routledge handbook of digital writing and rhetoric* (pp. 412–422). Routledge.

Haas, A., & Eble, M. (Eds.). (2018). *Key theoretical frameworks: Teaching technical communication in the twenty-first century.* Utah State University Press.

Hong, G. (2015). *Death beyond disavowal: The impossible politics of difference.* University of Minnesota Press.

Itchuaqiyag, C. U. (2021). Iñupiat ilitquisiat: An Indigenist ethics approach for working with marginalized knowledges in technical communication. In R. Walton, & G. Y. Agboka (Eds.), *Equipping technical communicators for social justice work: Theories, methodologies, and pedagogies* (pp. 33–48). Utah State University Press.

Jones, N. (2016). The technical communicator as advocate: Integrating a social justice approach in technical communication. *Journal of Technical Writing and Communication, 46*(3), 342–361. https://doi.org/10.1177/0047281616639472

Katz, C. (2022). Vagabond capitalism and the necessity of social reproduction. *Antipode: A Radical Journal of Geography, 33*(4), 709–728. https://doi.org/10.1111/1467-8330.00207

Kelly, C. (2013). Building bridges with accessible care: Disability studies, feminist care scholarship, and beyond. *Hypatia, 28*(4), 784–800. https://onlinelibrary.wiley.com/doi/abs/10.1111/j.1527-2001.2012.01310.x

Malatino, H. (2020). *Trans care.* Minnesota University Press.

McGinn Valley, T., Zander, M., Jacques, L., & Higgins, J. A. (2023). "The biggest problem with access": Provider reports of the effects of Wisconsin 2011 Act 217 medication abortion legislation. *Wisconsin Medical Journal, 122*(1), 15–20. https://wmjonline.org/wp-content/uploads/2023/122/1/15.pdf

National Latina Institute for Reproductive Health (NLIRH). (2010). *Advancing reproductive justice in immigrant communities: Promotoras/es de salud as a model.* https://www.latinainstitute.org/wp-content/uploads/2010/01/NLIRH-AdvancingRJ-ImmCommunities-Jan2010-1.pdf

Novotny, M., & De Hertogh, L. (2022). Amplifying rhetorics of reproductive justice within rhetorics of health and medicine. *Rhetoric of Health and Medicine Journal, 5*(4), 374–402. https://doi.org/10.5744/rhm.2022.5020

Novotny, M., & Opel, D. (2019). Situating care as feminist rhetorical action in two community-engaged health projects. *Peitho, 22*(1). https://cfshrc.org/article/situating-care-as-feminist-rhetorical-action-in-two-community-engaged-health-projects/

Piepzna-Samarasinha, L. (2018). *Care work: Dreaming disability justice.* Arsenal Pulp Press.

Powell, M., Levy, D., Riley Mukavetz, A., Brooks-Gillies, M., Novotny, M., & Fisch-Ferguson, J. (2014). Our story begins here: Constellating cultural rhetorics. *Enculturation,* https://www.enculturation.net/book/export/html/6096

Promotores de Salud and Health Promoter Programs. (n.d.). *Planned Parenthood of Wisconsin, Inc.* https://www.plannedparenthood.org/planned-parenthood-wisconsin/inc/education/promotores-de-salud-health-promoter-programs

Puig de la Bellacasa, M. (2017). *Matters of care speculative ethics in more than human worlds.* University of Minnesota Press.

Reproductive Justice. (n.d.) *SisterSong: Women of color reproductive justice collective,* https://www.sistersong.net/reproductive-justice

Ridolfo, J., & DeVoss, D. (2009, January 15). Composing for recomposing: Rhetorical velocity and delivery. *Kairos: A Journal of Rhetoric, Technology, and Pedagogy*, *13*(2). https://kairos.technorhetoric.net/13.2/topoi/ridolfo_devoss/intro.html

Riley Mukavetz, A. (2014). Towards a cultural rhetorics methodology: Making research matter with multi-generational women from the Little Traverse Bay Band *Rhetoric, Professional Communication and Globalization*, *5*(1), 108–125. https://docs.lib.purdue.edu/rpcg/vol5/iss1/6/

Ross, L., Roberts, L., Derkas, E., Peoples, W., & Bridgewater Toure, P., (Eds.). (2017). *Radical reproductive justice: Foundations, theories, practices, critique.* Feminist Press.

Silliman, J., Gerber Fried, M., Ross, L., & Gutiérrez, E., (Eds.). (2004). *Undivided rights: Women of color organize for reproductive justice.* Haymarket Books.

The New York Times (2022, June 24). The Dobbs v. Jackson decision, annotated. *The New York Times.* https://www.nytimes.com/interactive/2022/06/24/us/politics/supreme-court-dobbs-jackson-analysis-roe-wade.html

Tuhiwai Smith, L. (2012). *Decolonizing methodologies: Research and Indigenous peoples.* (2nd ed.). Zed Books Ltd.

Walton, R., Jones, N., & Moore, K. (2019). *Technical communication after the social justice turn: Building coalitions for action.* Routledge.

Wilson, S. (2008). *Research is ceremony: Indigenous research methods.* Fernwood Publishing.

15

A PEDAGOGY OF ETHICAL ENGAGEMENT

Preparing Students for Technical Communication in Communities

Carrie Grant

Introduction

Let's imagine a best-case scenario for a community-engaged design project. Decisions are driven by users' needs, as articulated by real users themselves and facilitated by trusted community organizers and advocates (Agboka, 2013; Carlson, 2022; Moore & Elliott, 2016; Salvo, 2001). Relationships are grounded in trust, working through balanced perspectives, aligned goals, and commitment (Bloom-Pojar, 2022; Grant, 2022). This is already a tall order full of complexities and potential pitfalls. Now add students.

Navigating community engagement work ethically is challenging regardless of format, but including students in the process adds an additional layer of complications to consider. How do we facilitate building students' community and cultural competencies simultaneously with their technical skills? How do we match scaffolding for students with partners' project timelines? Within the limited framework of a single semester course, there's a lot to juggle to achieve successful community-engaged design.

Since the social justice turn in technical communication, many pedagogical approaches have been shared to prepare students for advocacy and activism (Haas & Eble, 2018; Jones, 2016; Tham, 2021; Walton & Agboka, 2021). Critical analysis and public design scenarios can help hone students' social justice-mindedness for their future careers and civic lives. Design thinking naturally facilitates social justice approaches thanks to its emphasis on user advocacy. Theoretical frameworks and analytical tools help us map roads to more socially just technical communication.

DOI: 10.4324/9781003469995-19

However, we must mind the gap between theory and practice, as Rose reflects on her early work in technical communication with social justice interests:

> For social justice work to move into the realm of advocacy or activism, the work needs to be done in concert and in solidarity with organizations and community groups who can help enact the knowledge in a transformative way.
>
> *(Rose & Cardinal, 2021, p. 91)*

In other words, for our social justice work to have an impact, we need to connect it to the systems and people able to put it into action. Community-engaged design work gives us that capability, as community partnerships can extend our practice in advocating for individual users to be applied more broadly to address community needs. We need to expand our concept of serving the user to also serve the good of their community. Critique, reflection, and even design done in isolation from real-world implementation isn't enough.

So too for students: preparation for social justice work is important, but students won't actually be *doing* advocacy or activism until that work extends beyond the classroom. How better can we prepare students to become activist technical communicators than to guide and support them through their first live effort? Then, if we are going to put social justice work into action through community-engaged design projects with students, we ought to do it well, and ethically, or what's the point?

In this chapter, I offer strategies for a pedagogy of ethical engagement for navigating community-engaged design work with students. I'll also share an example course sequence, mapped onto the design thinking process, of a community-engaged graduate content strategy course. These strategies and experiences are a distillation from my decade of teaching community-engaged courses, working in university engagement offices, and particularly from co-managing a community grant and technical writing program—Towson University's GIVE project (for more information on GIVE, see Stuckey, 2019).

Designing the Community-Engaged Course or Student Project

There are two prevalent ways to do project-based community-engaged design work with students. One is through a community-engaged/service-learning course design, where project development is scaffolded by the instructor throughout the semester. The other is through internships and independent projects. Similar strategies are valuable to both formats as I'll describe below, though I've put more emphasis on course design considerations given the more rigid structure of a semester-long course.

Forging Partnerships

An ethical community engagement pedagogy begins with deciding who to partner with, and the politics of who gets to be a community partner is worthy of its own extended analysis. For the purposes of a starting point, as Zosha Stuckey and I explicate in "Localize, Prioritize, Decolonize: A Guide for Consciously Supporting POC-Led Organizations" (Stuckey & Grant, 2023), we advocate for a partnership approach that prioritizes local grassroots needs over organizations and institutions that already occupy positions of privilege and power. Partnership decisions are a small but meaningful point of pushback toward distributive justice, as through them we can challenge the dominance in the nonprofit sphere of white-led organizations that tend to pathologize the communities they serve.

In addition to considering the politics of partnerships, community engagement with students necessitates organizational capacity. Small organizations are very often trying to do too much with too few people and too little time and money. This reality makes them great candidates for partnership in terms of the potential to do good and serve the greatest need. However, it can simultaneously make them challenging candidates for partnership within the context of a classroom. The limitations of small organizations can lead to delayed communication, missing documentation, limited scheduling, and lack of feedback, all of which can wreak havoc on even the most forgiving of course calendars.

In naming these potential barriers, however, I don't mean to discourage partnerships with small organizations. Instead, I recommend trying to gauge an organization's capacity before formalizing a partnership. At GIVE, we do this through background research, initial inquiries, and meetings with new potential partners, in addition to taking recommendations and advice from our existing partners' networks. For instance, we tend to start with some basic Googling. Even if an organization has an underwhelming web presence, we can typically find clues about recent events they've held, grants they've won, or social media efforts they've made—these are all positive signs that the organization is ready to grow, and potentially to partner with a class. We then notice how quickly a potential partner responds to our initial interest email. We're generous with this, especially since we are strangers at this point, but we do take note. During an initial meeting, we compare the organization's current needs with the fit of our current courses and internships. We check whether someone would be available for around two meetings a semester with students during class time. We ask about organization timelines compared to our semester schedules, as well as communication preferences (sometimes texting is a simple solution to overwhelmed email inboxes!). Then we think about it, go with our gut, and see if the feeling is mutual. Sometimes it isn't, and that's okay.

Partnership formation also comes with the choice of whether to develop partnerships yourself as the instructor or to allow students to form their own. I have utilized both models, and in my experience, the risk of ethical compromises—particularly of students not following through on outcomes for partners—is much higher when students form their own partnerships. In my current teaching and intern management practice, I allow graduate students to bring in their own partners for secondary course projects, and at GIVE, Zosha and I seek intern input and interests for new organizations to partner with but ultimately make the call ourselves.

Setting Student and Partner Expectations

Service-learning fundamentals emphasize the importance of the memo of understanding (MOU) for laying out expectations from both sides of a partnership. It's easy to gloss over this step, but it does make a difference, and sometimes clarifying expectations requires more than a document.

In my experience, the most important expectations to set for a community partner are:

1 How much time do you expect from them (in terms of meetings, emails, feedback, and guidance)
2 The level and range of work they can expect from students (i.e. some projects will be more successful than others, and some may not be ready-to-use by the end of the class)
3 A plan for following up on work that doesn't get done or doesn't meet quality standards by the end of the class (more on this point later)

For students, I find that two major challenges persist, no matter how frequently or firmly I try to set expectations:

1 Student discomfort with the lack of certainty and ongoing fluctuation of community engagement compared to standard courses/projects
2 The level of quality needed for their work to be usable for the community partner, and the real stakes of their work compared to typical class assignments

Perhaps you'll perfect these struggles in ways I have not, but I believe the uncertainty and real stakes of community-engaged projects to be at the core of this work. What students will be writing depends on what the class learns from the community partner. Needs may evolve, and so deadlines may need to change. Compared to a typical class where the impact of students' work is merely their own grade, community-engaged design projects up the stakes with real-world implications.

Are there ways to make the process easier for students? Sure there are. I could create a strict schedule, pre-emptively formulate and divide project focuses, develop new examples for assignments, relax quality standards, and keep to the original course plan regardless of shifting partner needs. However, I find that such simplification for students' comfort is often to the detriment of partners' outcomes. I have seen student-centric versions of community engagement where students leave feeling positive about all the good they've done for the community, and community partners leave with very little they can actually use (Stoecker & Tryon, 2009). I've taught some of those classes myself. Ultimately, I believe they do not fulfill our responsibility to community engagement ethics or our commitment to socially just advocacy.

Rather than curating a rose-colored experience for students, asking them to grapple with uncertainty and real-world standards has the potential to teach them so much more than a patronizing pat on the back, building their adaptability and resilience for when, as we hope, they carry on their own social justice work in the future. The real-world constraints, compromises, and stakes of community-engaged design projects are their very value. So it's important that we are clear with students about what the process will look like and what we expect of their work. But it's also helpful to set our own expectations as instructors that some students will inevitably struggle with these expectations. Some will ultimately soar with them. Others will not. Maybe it will come to them later. Keep going anyway.

Foregrounding Intercultural Prep Work

So we've set up our community partnership and project, and we've done our best to make sure all parties understand the process and expectations. Now we get to design, right?! Not quite. Now, we need to prepare ourselves and our students for the intercultural work ahead so that we don't enter marginalized communities ill-equipped and cause unintended harm.

Community engagement principles like reciprocity, asset-based community development, and the community/individual right to self-determination offer accessible, practical guidelines to prime students for work with community partners (CTSA Task Force, 2011; Furco, 1996). From my experience in a variety of writing subfields and institutional engagement spaces, I've observed that technical communication specialists can at times be reluctant to draw from engagement scholarship across disciplines. But we don't need to reinvent the wheel here; service learning and community engagement pedagogy resources abound from folks who focus exclusively on this kind of work, and engagement best practices can only enhance our effectiveness in bringing our technical communication expertise into communities.

Within technical communication, the three Ps from Walton et al. (2019)—positionality, privilege, and power—offer a valuable lens to start from, as does Pouncil and Sanders (2022) model of the "work before." Both emphasize the importance of doing some self- and societal reflection before diving into action, so we can enter into community collaborations with the self-awareness to be able to check personal cultural assumptions, listen generously to understand community needs, and locate strategies for coalitional action.

In addition to the theoretical and identity prompts, in GIVE classes, we always introduce students to historical context about our local community. Where we're located just north of Baltimore, it's neither responsible nor fully possible to do work in/with the city without acknowledging its deep history of segregation and Black community disenfranchisement. It's worth the time to investigate a local community's social history. If that community or history is unfamiliar, we can investigate it together alongside our students. Understanding local context is critical to being able to navigate community collaboration respectfully rather than invasively.

Scaffolding these preparatory readings and conversations has its own extra challenge of timing, however. For most design projects, it isn't ideal to delay getting started on the actual design until late in the semester once we've had time to do all the reading first. So how much prep work should we do? I find that the sweet spot for having students meet the community partner for a full semester project sequence is about Week 3. I've made anywhere from Week 2 to Week 6 work. For some types of projects, even later may make sense. The things I always want to make sure we've talked about before that first meeting are: reciprocity, town/gown histories, and some introductory social critique relating to the issues the community is facing.

Finally, here are the biggest types of missteps I've seen students, faculty, and myself make that typically reflect a gap of some kind in the intercultural, historical, and personal reckoning work discussed above:

1 Asking partners condescending questions, like "Have you considered creating an Instagram?" (Answer: Yes, of course, they have. If they

haven't already done it, they either don't have time or have assessed that it's lower on their list of priorities than other, more pressing needs. Also, does their target audience even use Instagram?)

2 Pushing their own agenda rather than respecting the partner's expertise in their own community. For instance, insisting on the need for an app, based on a class focus or intern's technical skill, despite the partner or community members' feedback that they wouldn't utilize it.

3 Failing to recognize or honor legitimate community reservations about working with academic and other institutions. Just about any shade of "Why doesn't the community want to be studied?" can be answered by looking at the track record of negative consequences of communities working with universities in the past.

We of course want to try to prevent these potential breakdowns of trust within a community partnership. But even more important to sustaining ethical community engagement is how we repair after a misstep to make things right. If a student says something offensive in class, I interject in the moment if I can, or follow up with the partner privately to apologize. If I'm the one who has made the misstep, realizing perhaps I got a little too excited about a technical solution the community doesn't say they need, I recalibrate and listen again. Approaching partnership relationships with openness and humility is perhaps both more difficult and more important than any of the design work itself.

Iterating Feedback and Troubleshooting Technical Barriers

Once we've formed a partnership, set expectations, and prepared to engage ethically, we finally can dig into the design work itself. We've strategized how to divide our design work amongst student teams. We have our community partner facilitating access to and information about the community of our potential users. Then we hit our first technical roadblock or unexpected redirection.

Obstacles and iteration are inherent to technical design work, but community engagement can make them even trickier to work through. I've run into barriers like incompatibilities with platforms, difficulty clarifying information at critical junctures, surprise timeline changes, missed deadlines, blocked access to photos or data, subpar student drafts, late feedback scrapping an entire project, and more. Many of these barriers are really about planning and communication, as well as balancing accountability between the student and partner sides of the project.

I envision my role in the partnership as a facilitator, the go-between to help students and community partners understand each other, the project manager aiming to anticipate and mediate problems. When navigating a community project in flux, I'm always trying to creatively balance ways to serve both student and partner needs. For instance, say I realize that student drafts are not where I hoped them to be by our deadline to send them to the partner for feedback. Do I have the partner read them now anyway, even though I still have lots of feedback for students myself, and we may wind up needing more nuanced feedback from the partner again later? That feels like a potentially unfair drain on the partner's time. Instead, I'd do my own extra round of feedback for students and push back the partner feedback workshop by maybe a week if I can. Or from the other side, maybe a partner hasn't had time to read student drafts in full before our workshop meeting. Should I push back the workshop here too? Come down on the partner? Or, could I maybe highlight particular problem areas for us to read together live during our meeting, then allow the partner the opportunity to follow up later with full comments if they desire? (This is usually the option I choose.)

In both of these scenarios, I'm trying to think through how to best keep the project moving, as well as how to prevent shortfalls on either side of the project from derailing or overburdening the other side's work as much as possible. We're missing something from the community partner? Well, what else can students work on now, while I follow up with the partner? Students' work is missing the mark? How can I intervene before making this the partner's problem?

This balancing act is one of the hardest things about teaching community-engaged design, and something I think often prevents instructors from taking it on or drives them away after a bad experience. This negotiation never fully goes away, but we can build up our skills in anticipating setbacks and recalibrating them with resiliency. Scaffolding in project design is also key, building flexibility, revision, and buffer time where possible. I'll give an example of my setup for a community-engaged content strategy class in the next section.

Seeing Projects Through to Community Implementation

The final strategy for ethical community-engaged course and project design is one I feel is often forgotten from consideration. When does a community-engaged design project end? When the semester-long course ends? Not necessarily. I've written before about the importance of assessing community impact from partnerships, not just student outcomes (Grant, 2022). Following through on projects, even in light of potential shortfalls of student work, is important to meeting the ethical promise of reciprocity in community-engaged design.

The risk that too often comes to fruition is that a community-engaged design project ends with recommendations, a draft, or a prototype, and then the community partner, still overwhelmed with their daily demands and without the technical expertise that led them to seek support in the first place, never does anything with the project. Some would say that what the partner/client chooses to do or not do with your work is their prerogative, and that's fair enough. But, from what I've seen, the failure to implement a project is seldom the result of a clear, purposeful decision. Instead, the project likely was not ready to use when the partnership ended, and the barriers for the community partner to make it ready themselves were too great to overcome. And so the project sits, unused, going to waste.

Again, in this case, the biggest tool to prevent future problems is early planning. To increase the likelihood that student deliverables will be ready to implement, I build into our calendar multiple rounds of feedback and revision—typically first with comments from me, then feedback from users and/or the partner, and then a final check with me that the partner and user feedback is being addressed. I also build in extra time after a "final" deliverable is due in order to make further last-minute changes and execute the final product's launch.

Some projects are best planned from the beginning as ongoing, multi-semester efforts. We have many of these at GIVE, allowed by the structure of our ongoing core partnerships, rotation of faculty teaching different community-engaged classes, and regular access to student interns. Yet even with our long-term partnership model, we've sometimes abandoned projects. It happens. Sometimes the timing, circumstances, finances, or logistics just aren't right. Still, it's important to community engagement ethics to strive to fully see projects through as frequently as we can. Facilitating internships and independent studies is a particularly useful tool to adapt to unplanned additional project needs after a course ends. I have also been known to make final adjustments to student work between semesters and hit publish/print/send myself, though I'm in a constant process of trying to minimize the need to do so. If it's really not possible to facilitate a needed extension of the project, I recommend at least offering to brainstorm with the partner about possible routes for the project's completion. In order to do community-engaged design ethically, we owe our partners some accountability for the outcomes.

Scaffolding Assignments with the Design Thinking Process

Now I want to illustrate what a course structure might look like keeping these ethical strategies in mind. I'll also highlight how the design thinking process can map onto technical communication project scaffolding

(Grant, 2018; Tham, 2021). While the design thinking process alone doesn't necessarily result in social justice advocacy, it can be a valuable technical communication tool used in support of engagement ethics.

The assignment sequence below comes from a graduate professional writing course titled "Content Strategy and Writing for Social Change," and I structure my course loosely following Howard's (2020) model. For this particular semester, Fall 2023, I had planned in advance that the class would continue an ongoing GIVE partnership with the Northeast Towson Improvement Association (NeTIA), a slave-descendant community near Towson University's campus that is slowly being gentrified out of existence. The content strategy course would pick up where another graduate class had left off the previous semester, Halcyon Lawrence's "Technical Writing and Information Design" course, which had worked with NeTIA on a usability study of their current website, a redesign of an advocacy narrative white paper, and an interactive community map.

The expectations I set for my class's continuation of this work were that we would propose and implement updates to the website, building on the previous class's usability findings. We would also formulate a social and traditional media content strategy, complete with a batch of initial content to use, along with a plan for future GIVE interns to carry the strategy forward after the course (and additionally launch the map). The advocacy narrative white paper draft from Halcyon's class was already under revision by a GIVE graduate assistant who had taken the class, with the document on its way toward finalization and printing.

Design Thinking Step 1: Empathize—Community Background Work

With our content strategy class partnership in place and expectations for the semester laid out, we embarked on our first assignment: a competitive landscape analysis comparing NeTIA's existing content to other similar organizations. This is an internal class assignment intended to help students learn about our partner organization, envision its content possibilities, and hopefully work through any problematic assumptions about the partner before meeting them. While students began their background research, in class we discussed community engagement ethics in relation to personal identity, the organization's historical context, and introductory content strategy frameworks.

After this first project, the class went on a tour of the Northeast Towson neighborhood led by our community partner. Then we created proto-personas of NeTIA's target audiences. It would have been ideal for this step to include interviews with real members of each audience, which I have incorporated in previous semesters of the class. For this semester, I prioritized the community tour instead, considering we already had the site usability findings from the information design class. These are the kinds of scheduling compromises that must be weighed.

Even without the ideal of user interviews, this first phase of the course maps well with the "empathize" stage of the design thinking process, which is all about understanding users, and by extension, the community. Learning about our partner's exigence for wanting to amplify their community's visibility gave us critical context for later proposing culturally sensitive media campaigns that aligned with the organization's ethos. As several students commented after the neighborhood tour, they didn't fully get the significance of what NeTIA was trying to achieve until they heard the association's president's account of the histories and legacies of nearly every home in the neighborhood, dating back to their original settlement.

Unfortunately, not all students were able to make it to the tour, which I do believe limited their experience through the rest of the semester. As I mentioned with fostering students' intercultural competence, students also have to be willing to take the path you create for them, and not all will.

Design Thinking Step 2: Define—Content Audit

After students had a sense of the community and its needs, they conducted a comprehensive content audit of NeTIA's existing content and mapped user journeys. This was the transitional phase of analysis to identify the gaps between the organization's current strategies and their goals. In the design thinking process, the content audit is a strong fit for the "define" step, which is about identifying and clearly articulating users' needs. The user journey maps, developed based on the different personas navigating Ne TIA's website and other media, highlighted pain points where user needs were not being met by current content and design.

Once students conducted their content audit, they met again with Ne-TIA to present their findings and come to an agreement on goals for final strategy projects. We came to this meeting with multiple draft options of a core NeTIA content strategy statement to center our efforts around. Together, we landed on "NeTIA will shine a light on Historic East Towson through educational initiatives, community engagement, and advocacy for the residents and the land."

It's important to acknowledge here that this is a point in the semester where I had to begin balancing decisions between student and partner needs. Some audit groups' work felt underdeveloped by the time we had set for our partner presentation. Rather than delay in this phase and leave us with less time for workshopping the final deliverables, we proceeded with audit presentations as planned, and I put additional effort afterward into guiding particular groups.

Design Thinking Steps 3 & 4: Ideate and Prototype—Content Strategy Brainstorm and Draft

After the content audit and aligning our direction for content strategy projects with our partners, students got to work on the fun part: design brainstorming and drafting. Again this phase of the class aligns closely with the design thinking process steps, though I do mostly collapse steps three and four of ideating and prototyping. Students are encouraged to start by thinking creatively, and many brought forward insightful ideas that I would never have come up with on my own. A student with a background in graphic design created a brilliant new logo and branding guide for NeTIA. Others expanded the organization's toolkit for media engagement tenfold. Still more dove into an entirely unfamiliar web-hosting platform with significant usability limitations, and yet came out with a more accessible and intuitive site design. Each of these tasks came with its own set of technical and logistical barriers to work around and through.

At this phase, my ethical attention was focused on providing student feedback and support as frequently as I could, in the hopes of having strong products by the end of the term that we could actually launch. My success, as I'll explain, was mixed.

Design Thinking Step 5: Test—Community Partner Feedback

With a few weeks left in the semester, we met with NeTIA once more to present our "final" content strategy and deliverables. In the design thinking process, the next step is testing your design. Though this is traditionally the process's final step, it's meant to highlight the iterative nature of design—that once a product launches, it can throw you back into any of the previous steps to redefine new problems or refine solutions. This recursive philosophy suits community-engaged design work well.

For our "testing" phase, we presented our deliverables to NeTIA in as polished forms as we could, though I primed students with the expectation that we would have an additional round of revision in order to reach our partners' approval to launch the strategy projects. Our partners were effusive in their impressions of and gratitude for students' work. There were still adjustments to be made, as well as some projects where I pushed for further feedback based on my own assessment of the work, as well as my awareness that our partners might be hesitant to be overly critical even in areas where they were dissatisfied. Still, overall, the presentation of our content strategy for NeTIA was a clear success.

Design Thinking Step 6: Implement—Strategy Launch

Though as I mentioned, the design thinking steps typically stop at five, I'm following Tham's (2021) addition of a sixth step to emphasize implementation. After the strategy presentation, the class turned to final refinements and tying up loose ends. Their final launch results and follow-through vary.

The website team had some catching up to do, as the site's domain had actually gone down for about a week leading up to the presentation—talk about troubleshooting technical barriers! By the end of the class, we were able to work with our partners to get the updated website launched. For months after the course ended, however, a persistent oversight remained that the website's original address still brought users to the old site, though the new site is what appeared on Google. Eventually, we were able to connect with NeTIA's web administrator to redirect to the new site. I want to note that if this partnership had been just a single semester, I'm not sure if/when the fix would have happened.

The realm that was furthest behind at the audit phase remained the straggler at the end: social media. Despite multiple rounds of feedback and clarification, the students' final strategy and initial batch of posts still had issues with cultural sensitivity and design principles. In future semesters of the class, I intend to provide more design resources and asset-based language exercises, but still, these results can happen. Fortunately, this semester I prepared in advance for GIVE interns to pick up social media responsibilities for NeTIA after the course. During finals week, we were able to meet again with NeTIA to get account access, as well as revise enough of the grad students' posts to have them pre-scheduled through our winter break, then have GIVE interns resume developing and posting content in the spring. This is actually my best success to date with launching a social media strategy, which I find particularly challenging to both achieve usable quality results and support organizations to sustain long term.

In the end, we had strong outcomes from the class, both in terms of what students accomplished and what we were able to actually get implemented for our community partner. However, this isn't to say that the process or my own ethical navigating proceeded perfectly—it never does. What's most important to sustaining this work is the commitment to continue to strive to navigate it ethically.

Key Concept Review

Fostering a pedagogy of ethical engagement requires ongoing reflexivity in bringing attention to community impact while also building students' skills to create their own social justice impact in the future. Practically speaking, there are a number of steps we can take when facilitating community-engaged design projects to support more ethical engagement:

1 When forging partnerships, we can prioritize marginalized community organizations and groups, while simultaneously considering a potential partner's capacity for work with students.
2 From the beginning of a project, we can explicitly set expectations with both students and community partners about what will be asked of them, giving all parties the opportunity to opt-out.
3 Before bringing students into the community, we can support their development of intercultural awareness and communication skills. Even if we've done this well, we can anticipate missteps in building trust with partners and proactively work toward repair when they happen.
4 As projects evolve and roadblocks arise, we can adapt schedules and tasks to support the strongest possible outcomes for both students and community partners. Often this means being prepared for iterative rounds of feedback and technical troubleshooting interventions.
5 Finally, when a project is nearing its conclusion, we can work to ensure that the project is seen through to implementation in the community, even and especially if that means creating a plan to finish it out after a semester ends.

In mapping these strategies onto our course design, we can utilize the design thinking process to support project structure and social justice considerations:

1 Empathize: The first step of design thinking aligns well with the intercultural preparatory work students need to understand their community partners and potential users before beginning to design on their behalf.
2 Define: Once students have an initial understanding of the community, they can define the project at hand following the partner's guidance. What are the community's needs and challenges, and what are the parameters for how we will work to address them?
3 Ideate: With a deep understanding of the community's needs, we can now think creatively and expansively about possible design solutions

before narrowing our focus and dividing student tasks. Good ideas that are beyond the scope of this project can be saved for future opportunities.

4 Prototype: The hands-on drafting phase of community design projects is where technical and communication barriers inevitably arise. We can adapt to them by flexibly negotiating between students' and partners' needs.

5 Test: It's important to build in time for community partners and users to test and give feedback on a full working version of a project. This can defy students' mindset of "final" end-of-semester work, so it can be helpful to call a version "final" before it really is.

6 Implement: This is the step that takes student social justice work from practice to action through community-engaged design! Moreover, community partners often struggle to launch a project themselves if it's handed off incomplete at the end of a semester, so we should strive to see projects through to implementation.

Conclusion

Considering these steps for both navigating and structuring community-engaged design work with students, we can see how adopting a pedagogy of ethical engagement ultimately means bringing social justice to the forefront of not only our design purposes but also our design processes. By working with students on design projects that serve real community needs in partnership with grassroots community leaders, we can bring student preparation for social justice work to life, and practice what we preach in terms of the potential for advocacy and activism that technical communication affords. To lead students into community-engaged design work with integrity, we must not allow our commitment to student learning to compromise our commitment to preventing community harm in the process. Ethically, this is a never-ending balancing act, but one that's worthwhile for the growth and impact it can enable for students and communities alike.

References

Agboka, G. (2013). Participatory localization: A social justice approach to navigating unenfranchised/disenfranchised cultural sites. *Technical Communication Quarterly*, 22(1), 28–49. https://www.tandfonline.com/doi/abs/10.1080/10572252.2013.730966

Bloom-Pojar, R. (2022). Co-creating stories of Confianza. *Community Literacy Journal*, 17(1), 104–107. https://digitalcommons.fiu.edu/communityliteracy/vol17/iss1/11/

Carlson, E. B. (2022). "Who am I fighting for? Who am I accountable to?": Comradeship as a frame for nonprofit community work in technical communication. *Technical Communication Quarterly*, 32(2), 165–180. https://www.tandfonline.com/doi/abs/10.1080/10572252.2022.2085810

CTSA Task Force on the Principles of Community Engagement (2011). *Principles of community engagement* (2nd ed.). National Institutes of Health.

Furco, A. (1996). Service-learning: A balanced approach to experiential education. In B. Taylor & Corporation for National Service (Eds.), *Expanding boundaries: Serving and learning* (pp. 2–6). Corporation for National Service.

Grant, C. (2018). Tactics for connecting entrepreneurship and technical communication through community engagement: Experience report. In *Proceedings from the 36th ACM International Conference on Design of Communication* (pp. 1–6).

Grant, C. (2022). Collaborative tactics for equitable community partnerships toward social justice impact. *IEEE Transactions on Professional Communication*, 65(1), 151–163. https://ieeexplore.ieee.org/document/9712638

Haas, A. M., & Eble, M. f. (Eds.). (2018). *Key theoretical frameworks: Teaching technical communication in the twenty-first century*. Utah State University Press.

Howard, T. (2020). Teaching content strategy to graduate students with real clients. In G. Getto, J. T. Labriola, & S. Ruszkiewicz (Eds.), *Content strategy in technical communication* (pp. 119–153). Routledge.

Jones, N. N. (2016). The technical communicator as advocate: Integrating a social justice approach in technical communication. *Journal of Technical Writing and Communication*, 46(3), 342–361. https://journals.sagepub.com/doi/10.1177/0047281616639472

Moore, K. R., & Elliott, T. J. (2016). From participatory design to a listening infrastructure: A case of urban planning and participation. *Journal of Business and Technical Communication*, 30(1), 59–84. https://journals.sagepub.com/doi/10.1177/1050651915602294

Pouncil, F., & Sanders, N. (2022). The work before: A model for coalitional alliance toward Black futures in technical communication. *Technical Communication Quarterly*, 31(3), 283–297. https://www.tandfonline.com/doi/full/10.1080/10572252.2022.2069288

Rose, E. J., & Cardinal, A. (2021). Purpose and participation: Heuristics for planning, implementing, and reflecting on social justice work. In R. Walton, & G. Y. Agboka (Eds.), *Equipping technical communicators for social justice work: Theories, methodologies, and pedagogies* (pp. 76–97). Utah State University Press.

Salvo, M. J. (2001). Ethics of engagement: User-centered design and rhetorical methodology. *Technical Communication Quarterly*, 10(3), 273–290. https://www.tandfonline.com/doi/abs/10.1207/s15427625tcq1003_3

Stoecker, R., & Tryon, E. A. (Eds.). (2009). *The unheard voices: Community organizations and service learning*. Temple University Press.

Stuckey, Z. (2019). Grantwriting infrastructure for grassroots nonprofits: A case study and resource for attempting to 'return stolen things'. *Reflections: A Journal of Community-Engaged Writing and Rhetoric*, 19(2), 141–169.

Stuckey, Z., & Grant, C. (2023). Localize, prioritize, decolonize: A guide for consciously supporting POC-led organizations. *Open Words: Access and English Studies*, 14(1), 7–31. https://wac.colostate.edu/docs/openwords/v14n1/stuckey-grant.pdf

Tham, J. (2021). *Design thinking in technical communication: Solving problems through making and collaboration.* Routledge.

Walton, R., & Agboka, G. Y. (Eds.). (2021). *Equipping technical communicators for social justice work: Theories, methodologies, and pedagogies.* Utah State University Press.

Walton, R., Moore, K. R., & Jones, N. N. (2019). *Technical communication after the social justice turn: Building coalitions for action.* Routledge.

16

INVITING DISABILITY INTO THE TPC CLASSROOM THROUGH SERVICE LEARNING

Ellen Cecil-Lemkin

Introduction

During my doctoral program, I went from someone who could pass as nondisabled to someone who was marked by disability. This marking came about because I began using a service dog to manage my disability symptoms. My experience with disability began well before then, but during the last several years my symptoms had reached an intolerable level. I had felt a growing sense that I might be struggling to manage my symptoms unsuccessfully for years to come, when I learned that a service dog could be a solution. After researching organizations that could pair service dogs with disabled handlers, I quickly realized that service dog waitlists were years long and often required tens of thousands of dollars. Being a broke graduate student with persistent symptoms, I knew that these organizations weren't the answer. Luckily, I learned the Americans with Disabilities Act offered legal protection for disabled handlers to train their own service dogs. The legal protection didn't mean I was suddenly knowledgeable on how to train a service dog, so I sought the support of a qualified dog trainer. I was able to find a company that assists disabled individuals in training their dogs in service work. With my potential service dog candidate secured, I began working intensively and extensively with a trainer to prepare my dog for service work. Through this partnership, I developed a close relationship with the owners—one who is a disabled service dog handler themselves. When the owners shifted their focus to developing a nonprofit to breed and train service dogs, I became curious as to how my scholarship and teaching could leverage this opportunity to serve the service dog community.

DOI: 10.4324/9781003469995-20

As this was happening, I was also studying and teaching at Florida State University. The classes I taught were primarily in the Editing, Writing, and Media (EWM) major (The English Department, n.d.). Students in the major take classes that focus on developing their writing, composing, and critical thinking skills to develop textual and multimodal materials that cater to audiences' requirements in distinct contexts, a core value of technical and professional communication (TPC) (Zdenek, 2020, p. 538). One of the classes I taught in the major, Writing and Editing for Print and Online (WEPO), teaches students about the principles of composing and editing across different media environments, paying special attention to the constraints of each environment and the changes and challenges that occur as they work in and across each type. The overall goals of the course were to help students create and read texts differently, become informed about how audiences will interact with their texts, and bring intentionality to their composing and editing. Generally, there were four types of projects (print, digital, networked, and eportfolio) instructors were responsible for assigning, but the details, scaffolding, and scope were left up to the instructor to decide.

I had taught WEPO once before, and I felt that the initial iteration wasn't working quite well. While I had students composing and editing in different modalities, it lacked a central focus and student investment. Based on the feedback from the first class and talking to other instructors, I wanted my students to leave the course with compositions that could have lives and uses outside of the classroom. I decided to focus the class on professional communications they would use on the job market—cover letter, resume, eportfolio, and sample compositions to include in the eportfolio. To include a genuine, meaningful sample project, I created a service learning unit partnered with the nonprofit service dog organization I was connected with. This focus aligned me with scholars who have argued for forefronting disability in TPC courses, to consider it a central component for inclusion, rather than as a special additive (Heilig, 2023; Oswal & Palmer, 2022; Palmeri, 2006; Zdenek, 2019; Zdenek, 2020). After teaching this course three times, I saw how partnering with an organization for and led by disabled people increased student investment in accessibility and positively impacted the disabled community.

Based on this experience, this chapter explores the inclusion of disability in the TPC classroom through the use of a service learning project. While TPC scholarship has explored various pedagogical approaches to center disability (e.g., Browning & Cagle, 2017; Oswal & Palmer, 2022; Palmeri, 2006), it is yet to consider how service learning can deepen our students' investment in accessibility. To illustrate this point, I will first provide a rationale as to why this approach offers a timely intervention for the TPC field. Then, I will move into an explanation of how TPC instructors can implement a similar service learning approach into their courses. Through learning design principles that forefront disability and the lived experiences of disabled stakeholders, I'll demonstrate how instructors can prompt students to compose for, about, and with disability.

Toward Communal Accountability

In "Faceless students virtual places: Emergence and communal accountability in online classrooms," Kristie Fleckenstein (2005) addresses the impact of "facelessness" in online courses. Facelessness is used to describe how students enrolled in online courses struggle to imagine real people attached to names and comments and the consequences of their (in)actions. Fleckenstein (2005) identifies the lack of physical presence as a significant contributor to facelessness. As a result, students "behave irresponsibly towards each other and toward the ambient community because neither is perceived as real" (Fleckenstein, 2005, p. 151). The difficulty of facelessness persists beyond students interacting in an online environment. It is a challenge that students grapple with whenever they're asked to compose for imagined (or "faceless") audiences. Students wrestle with valuing audiences they cannot connect to or imagine as real. Facelessness persists when we ask students to compose for disabled audiences.

This is not only a conundrum for students, but a matter the field struggles with as well. There's been over 20 years of scholarship in TPC asking for increased focus on accessibility and disability (Heilig, 2023), and yet, there's limited scholarship addressing this area and continued resistance to centering accessibility. I sense that accessibility and disability inclusion might seem like more effort than reward because these concepts appear immaterial and tied to imagined people. Indeed, even within our own publications, much of the disability TPC scholarship is written by authors who do not identify as disabled or acknowledge their disability status, which contributes to the othering and abstraction of disabled people (Linton, 1998; Price, 2012). If the material and emotional impact (in)action has on disabled people is abstract, we do not see the consequences of our (in)actions impacting real people. It then becomes easy to deprioritize the adoption and education of accessible design practices.

The facelessness of disabled people contributes to both the TPC field and classroom lacking communal accountability or sense of interdependence. Fleckenstein (2005) defines communal accountability as:

> ...the reciprocal commitment among individuals to act in ways that promote the evolution and health of their interconnections. Such

answerability requires people to recognize the value of group cohesiveness, their inherent interdependence, and their individual responsibility for the well-being of one another.

(p. 150)

In other words, individuals act to support community members because it impacts the community's and their own well-being. As it currently stands, the TPC classroom and field is lacking a sense of communal accountability for disabled scholars and users. When our (in)actions continue to perpetuate ableism or access barriers, we fail to imagine our interdependence. To foster communal accountability, we need to create paths toward "grounding" disabled people as real—not imagined or hypothetical—people.

One way to move toward this grounding is through the centering of disabled people's narratives. Kristen Moore (2013) has drawn TPC's attention toward the value of storytelling for building students' capacity for relational work—"work that draws attention to the complex relationships among people, ideas, places, events, institutions, and things" (p. 63). Storytelling's ability to expand our capacity for relational work is not limited to the classroom. This technique can also be used to strengthen the field's relational work and create a grounding for communal accountability. To this end, there has been some progress, most notably in rhetoric and composition. In 2020, *College Composition and Communication* released a symposium on the lack of accessibility in conference spaces (Hubrig & Osorio, 2020). The symposium consisted of articles written by disabled people navigating inaccessible conferences and the negative impact of these access barriers. Their narratives illuminate the effects of our (in)action on real people. With this goal in mind, I used a narrative framing for my introduction to this chapter.

While narratives provide an important step in identifying real people within the field and the consequences of inaccessibility, they can still leave disabled people faceless. Work is needed that goes beyond the page to engage with disabled people. TPC scholars have called for this work through participatory design (Oswal, 2014; Oswal & Palmer, 2022). Participatory design centers on disabled users by collaborating with them and incorporating their feedback into the design process (Oswal & Palmer, 2022, p. 251). Sushil K. Oswal and Palmer (2022) argue that this work is "essential for our field to remove its veil of disability ignorance and experience the value of disability first hand" (p. 252). Participatory design can offer us an approach to accessible design and pedagogy that is grounded in the realities of disabled people.

Toward this goal, participatory design offers an intervention in the abstraction of universal design. The goal of universal design is to create spaces and objects that are accessible to as wide a range of users as possible, without the need for retrofitting (Hamraie, 2013). While universal design is a worthy goal, it also contributes to a sense of groundlessness since its goal can never truly be realized due to the range of disabilities and competing access needs. Universal design presents us with a direction to move in, but it doesn't offer an endpoint—leaving us faltering and dissatisfied. Participatory design in conjunction with universal design offers us a movement toward accessibility that is grounded in the realities of the disabled people it's intended to serve. The nature of this combined design approach will lessen the need for retrofits (Oswal, 2014, p. 15), signal our value for disabled users, and fuel sustained accessibility work.

Within the classroom, creating opportunities for participatory design offers instructors and students avenues to anchor their work within the lived experiences of disabled people. I see service learning as a method to develop communal accountability for accessibility through participatory design. Service learning is broadly understood as "a cross-disciplinary program whose main focus is pedagogy and service" and "a method of teaching multiple skills" (Adler-Kassner et al., 1997, p. 2). In many service-learning courses, a class pairs with a community partner or an organization to support its mission in ways that are aligned with course learning objectives, which creates a mutually beneficial outcome for students and the organization. Scholarship on service learning has called for instructors to be active participants in these partnerships to build trust, sustainability, and meaningful work for the organizations (Cushman, 2002). In many ways, ethical service learning that avoids the "hit-it-and-quit-it relation with communities" (Cushman, 2002, p. 41) draws upon participatory design principles. Pairing service learning with participatory design offers a complimentary approach to center disabled people and bolster investment in accessibility.

While I'm not claiming that this approach will result in a radical transformation of the field and classroom, I do think it can take us closer to creating a culture of access grounded in access intimacy. Disability activist Mia Mingus (2017) explains:

> Access intimacy at once recognizes and understands the relational and human quality of access, while simultaneously deepening the relationships involved. It moves the work of access out of the realm of only logistics and into the realm of relationships and understanding disabled people as humans, not burdens.
>
> *(para. 18)*

I understand access intimacy as a deeply empathetic approach to accessibility that centers on the inherent value of disabled people. It requires us to see disabled people as real and accessibility "as an act of love" (Mingus, 2018, para. 34). Without access intimacy, we run the risk of undercutting the value of accessibility, retrofitting accessibility, seeing accessibility as a static checklist, or adopting accessibility practices without challenging the ableist assumptions that necessitate them. Working with disabled people in service learning classrooms can move us in this direction.

To assist instructors in developing a curriculum that deepens students' investment in accessibility through service learning, I offer three modules and course preparatory considerations to scaffold student knowledge. The subsequent modules outline a rationale for why these concepts should be addressed and include an outline for specific topics. In addition to these module descriptions, I've also included a handout at the end of this chapter. This handout offers recommendations for additional resources that instructors can use to deepen their understanding of the concepts or assign to their students as homework. Each module can be taught in one to two class sessions, depending on the instructor's preferences and the content needed to be covered. Additionally, while I've imagined the service learning work to be one multiweek unit of a class, this approach could be expanded to scale the entire semester.

Preparing to Teach the Service Learning Unit

The work of centering accessibility and fostering access intimacy begins during the planning stage of course development. Instructors who want to demonstrate their value for and commitment to accessibility should design their classes using the accessibility principles they plan on teaching. Teaching these principles without implementing them into the course signals a lack of genuine investment and care for disability and accessibility. Beyond demonstrating their commitment, there's a great chance that disabled students enrolled will be enrolled in those classes. Working with a disabled community partner should not lead us to forget that disability exists inside the classroom too. The number of disabled students in higher education is growing, with the current national numbers at 20.5% (National Center for Education Statistics, 2023). Designing an accessible class from the beginning grounds instructors in the realities of their disabled students and works toward mitigating the instructor's perception of facelessness of disabled students. This work, in turn, can foster access to intimacy and contribute to a welcoming and inclusive classroom and environment for disabled students.

Part of the early stages of course design should also include identifying potential partners to work with. For those instructors without connections developed from lived experience, the process of identifying and establishing a relationship may take time. In part because the potential partner is agreeing to a time commitment themselves in order to enact participatory design effectively. Furthermore, instructors need to be intentional in their selection, being careful to identify organizations that are both for and led by disabled people. Disability activists have long called for "nothing about us without us" and more recently "nothing without us" (Crowther, 2007). Organizations without disabled leaders run the risk of "perpetuating structural inequalities at the cost of further marginalizing the recipient" (Oswal & Palmer, 2022, p. 257).

Module 1: Disability Basics

I begin the service learning unit on disability basics to provide students with a foundation to understand and interrogate concepts of disability and accessibility. Many students have had little exposure to topics related to disability during their education. I have found that undergraduate students tend to have a limited understanding of disability, which is often connected to what they've seen in media representations. While there are always exceptions, I have frequently encountered undergraduate students with disabilities who would not describe themselves as disabled due to misconceptions about disability. Introducing students to the basic concepts around disability will expand their understanding of disability and create a framework for considering access for both the unit and their future projects. To jumpstart this work, I recommend addressing these three ideas:

- prevalence of disability
- definitions of disability
- the social vs. the medical model of disability

Addressing the prevalence of disability in the national context is important, since many students are unaware of how large the disabled population is. For example, in the United States, disabled people make up 27% of the population, making us the largest minority (CDC, 2024). Conversations around disability prevalence help to solidify the importance and impact accessibility can have on a public audience. It also leads to questions on what counts as a disability. Disability is one of those tricky terms that defies easy definition. The different definitions often reveal nuances and complexities, showcasing how it resists simple classification and languageing. Students are usually surprised about the scope of the definitions and the knowledge that their conditions are considered disabilities. Showing definitions by the

Americans with Disability Act, United Nations, and *Keywords for Disability Studies* can provoke critical thought on these variations. Finally, when introduced to the medical and social models of disability, students can then draw on a framework to analyze accessibility and disability. While these models have been critiqued, they also provide a helpful beginner framework to sift through these concepts. As Browning and Cagle (2017) have stated, "the social model is a useful entry into what for many students is a revolutionary theoretical framework" (p. 451).

Module 2: Disability Specifics

The second module goes beyond the basics to offer some specific disability knowledge directly related to the service learning partner's community. The disability overview offered in the previous module will provide general disability knowledge, but it will be too broad to offer individual guidance related to a specific disability experience or context. Students need specific knowledge related to the organization you're partnering with in order to be conscious and responsive designers. By focusing on the particulars of the disabled community, students will be better prepared to contribute meaningful designs in a way that is both respectful and responsive to them. Therefore, students will need some specialized information to help them better understand the need for the partner organization and the everyday experience of the disabled people it's serving. While the details will depend upon the context of the partner organization, some areas to address include:

- disability experiences and/or culture
- common access barriers
- service(s) offered by the partner organization

The lived experiences of someone with a particular disability vary widely from person to person. In other words, no two disabled people's experiences are exactly the same, regardless of their shared diagnosis. Therefore, information about specific disabilities should be framed as contextual and varied based on the individual. That being said, there can be enough commonalities to prove useful and prepare them for design considerations and conversations with the partner organization. Depending on the disability, it might also be relevant to include information about the specific disability community and its culture. For example, in my own class, I talked about common service dog etiquette and expectations. If the instructor is not a member of this particular disability community, there could be a large learning curve for them as well. My recommendation is to check with the partner organization to see what they want students to know before meeting with them. They may have recommendations for readings or other media for students. These supplemental materials are not to take the place of interacting and meeting with disabled people, but provide them with some foundational pieces that may lessen access fatigue—"the everyday pattern of constantly needing to help others participate in access, a demand so taxing and so relentless that, at times, it makes access simply not worth the effort" (Konrad, 2021, p. 180). Constantly requiring disabled people to educate others on their access needs is a taxing experience. The work of creating a culture of accessibility requires everyone to participate (Brewer et al., 2014; Hubrig & Osorio, 2020; Mingus, 2011), and should not be left to our disabled partners to undertake alone. Once that foundational knowledge is in place, the next step is turning toward the partner organization and their work. Students can review the social media or website of the organization to prepare themselves for meeting with the partner. A first meeting with the entire class and the partner can ensure that everyone has the same information and lessen the partner's time commitment. Instructors can prep students for a meaningful first meeting by brainstorming questions the group has about the community, common access barriers, the organization's mission, and the type of services provided.

Module 3: Designing for Access

The third module teaches students how to design accessible compositions. This module should take place after the class has had the opportunity to meet with the partner and identify some areas the class can support through their work. With the first two modules, students grappled with disability and began to understand the practical and effective reasons for accessibility. While they may now be invested in this work, they lack the knowledge and skill to enact it in a meaningful way. Designs created without knowledge of accessible design practices can lead to unusable projects that alienate the intended audience and community partner. For example, if the community partner serves d/Deaf people, creating multimodal projects that utilize sound without captions or a transcript would result in a product that excludes its intended audience. To prepare students to create accessible designs, this module should address:

- universal design
- participatory design
- specific accessibility components related to the students' compositions

As previously mentioned, universal design does have some limitations. Much like introducing the social and medical models, universal design offers students a theoretical framework to begin conceptualizing their access goals. It provides students with a direction to move in and a preliminary method to evaluate their own progress. The lack of definitive steps in universal design, necessitates the inclusion of participatory design. The work of the previous two modules lays the foundation for students to value participatory design by highlighting the limits of their knowledge and the need for disabled users' expertise. Participatory design offers students a place to ground their design and receive feedback. As students embark on participatory design, they also need practical information regarding the common accessibility features for their potential designs. For example, if they will be creating compositions with images, they'll need to be instructed in how to write and embed alt text and image descriptions. While there's been much written on avoiding a checklist approach to accessibility (Dolmage, 2015; Oswal & Melonçon, 2017; Zdenek, 2020), students (and, indeed all composers) need instruction on standard accessibility practices in order to implement them and avoid common access barriers. The trap of checklists, as I see it, is an overconfidence in their comprehensiveness, a flattened understanding of disability, and a refusal to engage in participatory design processes. If this approach is paired with participatory design, then students may be less likely to stumble into these pitfalls.

Key Takeaways and Design Activities

TPC has begun the work of fostering a culture of accessibility; however, there is still a tendency to deprioritize accessibility and disability inclusion. I have argued that the field's de-emphasis is the result of the "facelessness" (Fleckenstein, 2005) of disabled audiences and scholars. To make a turn toward consistent, empathetic inclusion of disabled perspectives and people, toward access intimacy (Mingus, 2017), we need to find ways to connect with the disabled people our work impacts. Without access intimacy, we run the risk of undercutting the value of accessibility, retrofitting accessibility, seeing accessibility as a static checklist, or adopting accessibility practices without challenging the ableist assumptions that necessitate them. Partnering with disabled people for service learning projects in our classrooms that draw from accessibility design principles of universal design and participatory design can help cultivate access intimacy. To begin this work, instructors must develop courses using the same design principles and facilitate genuine, meaningful relationships with nonprofit organizations that are for and led by disabled

people. When teaching the service learning unit, three modules addressing disability basics, disability specifics, and accessible design principles
should be taught. The first module on disability basics should include
information on defining and framing disability. The next module, disability specifics, should dive into the particulars of the disabled community
the students will be serving. The last module should cover the knowledge
and skills necessary to create accessible designs. Adopting this curriculum, working with real, non-imagined disabled people, and witnessing
the consequences of inaccessibility can deepen students' investment in accessibility and build access intimacy.

To promote students' engagement and comprehension of this material,
I offer these design activities that instructors could implement throughout
this unit:

- Mocking up sample multimodal designs for students to review and collectively identify potential access barriers;
- Providing sample images for students to practice writing alt text and
  image descriptions;
- Creating text documents with limited formatting for students to practice formatting changes to enhance visual readability and screen reader
  compatibility;
- Visiting locations on campus to discuss the apparent points of access
  and accessibility barriers in the physical designs and layouts;
- Workshopping drafts in progress to provide feedback on areas of
  strength and potential access barriers;
- Meeting with the community partner to receive feedback and guidance
  on the accessibility and usability of the designs.

Conclusion

My goal in developing a service learning unit was to strengthen my students' understanding of disability and capability to design accessible
compositions. What I've come to appreciate is its capacity to positively
influence students' commitment to accessibility. When TPC instructors
center disability by incorporating the stories and lived experiences of disabled people, teaching accessible design techniques, and enacting participatory design we can foster a deeper appreciation of the positive impact of
accessibility for disabled people. This goal, however, can only be accomplished if there's an authentic commitment by the instructor. With all of
the competing curricular goals and pedagogical considerations for TPC
courses, it can be all too easy to leave disability out of our curriculum
entirely. Partnering with disabled people through service learning can help

fortify our own dedication to the ongoing work needed for a culture of accessibility and access intimacy.

Resources for Instructors

Module 1: Disability Basics

Adams, R., Reiss, B., & Serlin, D. (Eds.). (2015). *Keywords for Disability Studies*. New York University Press.

Lebrecht, J., & Newnham, N. (Directors). (2020). *Crip Camp: A Disability Revolution* [Documentary]. Netflix.

Linton, S. (1998). *Claiming disability: Knowledge and identity*. New York University Press.

Sins Invalid. (2019). *Skin, Tooth, and Book: The Basis of Movement is Our People* (2nd ed.).

Module 2: Disability Specifics

Annie Elainey. (2016, February 4). *How to spot a fake disability* [Video]. YouTube. https://www.youtube.com/watch?v=sIOcKpSVp4k

Connor, D. J. (2008). *Urban narratives: Portraits in progress, life at the intersections of learning disability, race, & social class*. Peter Lang.

Imani. (2019, July 11). #WhenICallMyselfDisabled, your opinion doesn't matter. *Crutches and Spice*. https://crutchesandspice.com/2019/07/11/whenicallmyselfdisabled-your-opinion-doesnt-matter/

Wong, A. (Ed.). (2020). *Disability visibility: First-person stories from the Twenty-first century*. Vintage Books.

Module 3: Designing for Access

Dolmage, J. (2015). Universal design: Places to start. *Disability Studies Quarterly*, *35*(2), Article 2. https://doi.org/10.18061/dsq.v35i2.4632

Hamraie, A. (2013). Designing collective access: A feminist disability theory of universal design. *Disability Studies Quarterly*, *33*(4). https://dsq-sds.org/article/view/3871/3411

Oswal, S. K., & Palmer, Z. B. (2022). A critique of disability and accessibility research in technical communication through the models of emancipatory disability research paradigm and participatory scholarship. In J. Schreiber & L. Melonçon (Eds.), *Assembling critical components: A framework for sustaining technical and professional communication* (pp. 243–267). The WAC Clearinghouse; University Press of Colorado. https://doi.org/10.37514/TPC-B.2022.1381.2.09

References

Adler-Kassner, L., Crooks, R., & Watters, A. (1997). Service-learning and composition at the crossroads. In L. Adler-Kassner, R. Crooks, & A. Watters (Eds.), *Writing the community: Concepts and models for service-learning in composition* (pp. 1–18). American Association for Higher Education.

Brewer, E., Selfe, C. L., & Yergeau, M. (2014). Creating a culture of access in composition studies. *Composition Studies, 42*(2), 151–154.

Browning, E. R., & Cagle, L. E. (2017). Teaching a "critical accessibility case study": Developing disability studies curricula for the technical communication classroom. *Journal of Technical Writing and Communication, 47*(4), 440–463. https://doi.org/10.1177/0047281616646750

CDC Newsroom. (2024). CDC data shows over 70 million U.S. adults reported having a disability. https://www.cdc.gov/media/releases/2024/s0716-Adult-disability.html

Crowther, N. (2007). Nothing without us or nothing about us? 1. *Disability & Society, 22*(7), 791–794. https://doi.org/10.1080/09687590701659642

Cushman, E. (2002). Sustainable service learning programs. *College Composition and Communication, 54*(1), 40–65. https://doi.org/10.2307/1512101

Dolmage, J. (2015). Universal design: Places to start. *Disability Studies Quarterly, 35*(2), Article 2. https://doi.org/10.18061/dsq.v35i2.4632

Fleckenstein, K. S. (2005). Faceless students, virtual places: Emergence and communal accountability in online classrooms. *Computers and Composition, 22*(2), 149–176. https://doi.org/10.1016/j.compcom.2005.02.003

Hamraie, A. (2013). Designing collective access: A feminist disability theory of universal design. *Disability Studies Quarterly, 33*(4). https://dsq-sds.org/article/view/3871/3411

Heilig, L. (2023). Critical disability studies in technical communication: A 25-year history and the future of accessibility. In M. S. Jeffress, J. M. Cypher, J. Ferries, & J.-A. Scott-Pollock (Eds.), *The Palgrave handbook of disability and communication* (pp. 401–415). https://doi.org/10.1007/978-3-031-14447-9_24

Hubrig, A., & Osorio, R. (2020). Enacting a culture of access in our conference spaces. *College Composition and Communication, 72*(1), 87–96.

Konrad, A. M. (2021). Access fatigue: The rhetorical work of disability in everyday life. *College English, 83*(3), 179–199. https://doi.org/10.58680/ce202131093

Linton, S. (1998). *Claiming disability: Knowledge and identity.* New York University Press.

Mingus, M. (2011, February 12). Changing the framework: Disability justice. *Leaving Evidence.* https://leavingevidence.wordpress.com/2011/02/12/changing-the-framework-disability-justice/

Mingus, M. (2017, April 12). Access intimacy, interdependence and disability justice. *Leaving Evidence.* https://leavingevidence.wordpress.com/2017/04/12/access-intimacy-interdependence-and-disability-justice/

Mingus, M. (2018, November 3). "Disability justice" is simply another term for love. *Leaving Evidence.* https://leavingevidence.wordpress.com/2018/11/03/disability-justice-is-simply-another-term-for-love/

Moore, K. (2013). Exposing hidden relations: Storytelling, pedagogy, and the study of policy. *Journal of Technical Writing and Communication, 43*(1), 63–78. https://doi.org/10.2190/TW.43.1.d

National Center for Education Statistics. (2023). *Students with disabilities.* Digest of Education Statistics. National Center for Education Statistics. https://nces.ed.gov/fastfacts/display.asp?id=60

Oswal, S. K. (2014). Participatory design: Barriers and possibilities. *Communication Design Quarterly, 2*(3), 14–19. https://doi.org/10.1145/2644448.2644452

Oswal, S. K., & Melonçon, L. (2017). Saying no to the checklist: Shifting from an ideology of normalcy to an ideology of inclusion in online writing instruction. *Writing Program Administration, 40*(3), 61–77.

Oswal, S. K., & Palmer, Z. B. (2022). A critique of disability and accessibility research in technical communication through the models of emancipatory disability research paradigm and participatory scholarship. In J. Schreiber & L. Melonçon (Eds.), *Assembling critical components: A framework for sustaining technical and professional communication* (pp. 243–267). The WAC Clearinghouse; University Press of Colorado. https://doi.org/10.37514/TPC-B.2022.1381.2.09

Palmeri, J. (2006). Disability studies, cultural analysis, and the critical practice of technical communication pedagogy. *Technical Communication Quarterly, 15*(1), 49–65. https://doi.org/10.1207/s15427625tcq1501_5

Price, M. (2011). *Mad at school: Rhetorics of mental disability and academic life.* University of Michigan Press.

The English Department. (n.d.). Editing, writing and media. Retrieved January 3, 2024, from https://english.fsu.edu/programs/editing-writing-and-media

Zdenek, S. (2019). Guest editor's introduction: Reimagining disability and accessibility in technical and professional communication. *Communication Design Quarterly, 6*(4), 4–11. https://doi.org/10.1145/3309589.3309590

Zdenek, S. (2020). Transforming access and inclusion in composition studies and technical communication. *College English, 82*(5), 536–544.

17

PEDAGOGICAL APPROACHES TO NORMALIZE INCLUSIVE DESIGN

Jamal-Jared Alexander, Dorcas A. Anabire, and Rebecca Walton

Introduction

When introducing students to professional design practices, we argue that technical communication instructors should aim to normalize inclusive design, encouraging students to habitualize *inclusive* practice as *professional* practice. When we say inclusive design, we're referring to students investigating ways to challenge the status quo by intentionally centering equity and justice throughout their design process (Iyamah, 2023). In other words, inclusive design encourages students to think about how people with different intersections of identity may experience or engage with the designs they make.

This chapter describes how we have integrated inclusive design into an undergraduate Visual Communication Design course that is structured around the design thinking process. We describe the background for this course design, then overview the course design, followed by describing two customizing options that have been implemented at two institutions. These options demonstrate that aspects of the course design can be adapted and modified based on available resources and networks. In fact, it's important to note that integrating inclusive design into technical and professional communication (TPC) courses is not a one-size-fits-all approach. Therefore, we ask readers to think of our customizing options as a starting place to help you enact your own version of designing for social justice.

DOI: 10.4324/9781003469995-21

Literature Review

Design Thinking Pedagogy and Social Justice

Cook's (2002) Layered Literacies framework offered a structure for technical communication curriculum that accounts for a wide range of literacies students should develop in the classroom that would prepare them to enter the workforce. Though this study was published more than two decades ago, it remains true that "workplace writers [and designers] need a repertoire of complex and interrelated skills to be successful" (p. 7). We argue that these skills have expanded to include complex considerations of designing for diverse populations, and this chapter conveys some of the ways in which we try to normalize inclusive design practices and considerations in an undergraduate course structured around the design thinking process.

Design thinking pedagogy, therefore, becomes crucial in considering inclusive design in our approach to teaching. Tham (2021) argues that though there has been increased attention to adopting design thinking principles in TPC classrooms, more is needed. Using a TPC course as a case study, Tham (2021) outlined pedagogical heuristics to help instructors practice design thinking in their pedagogy:

(1) making and design thinking in TPC pedagogy should advocate for an iterative design process, (2) making and design thinking should encourage students to pursue actionable solutions, (3) making and design thinking should leverage the always-already interdisciplinary and collaborative nature of TPC and promote open spaces for collaboration and (4) making and design thinking should encourage students to focus on human-centered design.

(pp. 403–404)

These heuristics are reflected in the revised version of the Visual Communication Design course discussed in this chapter. Bay et al. (2018), as cited in Tham (2021), posit that design thinking would help TPC instructors introduce concepts like user-centered design, usability, and collaboration, which were central to our design course. Moeggenberg and Walton (2019) argued that design thinking can increase students' motivation because design thinking offers students a process for applying problem-solving skills that they can utilize in their future workplace as well as "prepare students for transdisciplinary work environments and the kinds of wicked problems they are likely to encounter there" (para. 7). Lane (2018) offers an additional benefit of design thinking in the classroom: TPC instructors can use design thinking practices to help incorporate social justice into their pedagogy and pedagogical materials.

Moeggenberg and Walton (2019) describe one way to do so, by incorporating concepts from queer rhetorics into design thinking: working closets, queering time, and queering spaces. The claim and findings by Moeggenberg and Walton (2019) show that design thinking pedagogy can be approached in different ways to normalize inclusive design practices. In a similar claim, Tham et al. (2022) argue that design thinking "fosters 'radical' collaboration and democratization in TPC processes" (p. 452). We believe that when design thinking approaches are used in a service-learning course, it helps students involve their clients from the ideation stage of a project. Also, design thinking allows students to design for the specific needs of their clients, a consideration at the core of user experience (UX), broadly speaking.

UX instructors are known to customize their UX pedagogy to "document the complexities of their own UX teaching, offering bespoke approaches to those issues, and requesting more systematic scholarship for best practices" (Turner & Rose, 2022, p. 68). For example, Alexander customized the Visual Communication Design course to reflect his stance on service learning by partnering with university partners that serve marginalized student communities. Turner and Rose (2022) showed how instructors use a "complex set of pedagogical practices such as client-based projects, audience analysis, design, iteration, and usability evaluation" when they teach UX in TPC classrooms (p. 95). Lane (2021) proposes that enacting interstitial design, a concept that "considers the forming or occupying of interstices" (p. 217), will enable students to apply TPC "theories and practical skills (such as document design best practices, audience analysis, and recursive composing) to unique scenarios that illustrate how TPC principles can be applied to local problems" (p. 227). Although the literature demonstrates a range of ways to teach design thinking, we note a common thread throughout the scholarship discussed above: Design thinking pedagogy can help TPC instructors incorporate design theories, iterative design, problem-solving, and client-based design into their classrooms to achieve social justice and inclusive design.

Community Engagement, Service Learning, and Social Justice

The Visual Communication Design course discussed throughout this chapter is a service-learning course taught at our respective institutions. Service-learning, a "popular pedagogical tool used to teach students both social consciousness and pragmatic, real-world writing skills" (Stone, 2000, p. 385), is an essential part of the course, offering students a context for applying workplace design skills and connecting them to real clients that they can work with to produce documents by the end of the course. One key feature of service-learning is collaboration; it can be a great way to train students for the kind of collaborative writing and design they will be doing in the workplace (Morgan et al., 1987). Stone (2000) asserts that service-learning empowers student designers "as agents in the discourse of their neighborhoods and cities" (p. 397). Many service-learning courses have students work with clients who are local in the university community; in that sense, students are not only working to produce accessible documents for their clients but also contributing to the development of their community. This approach aligns with Hutter and Lawrence's (2018) claim that TPC classrooms "can be a space to think critically about accessibility and implications for great inclusiveness of users with a range of abilities" (p. 29). They further argue that applying usability testing techniques, such as card sorting,[1] can help students understand their users' needs, experiences, skills, and behaviors (Hutter & Lawrence, 2018). In other words, design thinking practices are central to UX because students get the chance to collaborate with their clients throughout the design process, reducing the chance of product inaccessibility.

Despite TPC embracing service-learning courses to help students transition to their workplace, there are challenges with engaging in service-learning. For example, Bourelle (2014) mentions that locating community partners and facilitating communication between students and clients are common challenges. For service-learning partnerships to be successful, clients must be actively involved in the projects; however, when communication with clients is subpar, students can become frustrated in the process and might not be able to accomplish assigned tasks.

Course Design: Vision for Social Justice

Background

Originally titled "Document Design and Graphics," this upper-division undergraduate course was one of the first classes Walton taught upon being hired at Utah State University (USU) in 2011. Initially, the course addressed topics such as design principles, color theory, and typography, with ethics explicitly informing lessons on data displays but little else. Course materials emphasized iterative design and audience analysis, but design thinking, as a formal process, was not part of the course. The class was taught in both the fall and spring semesters, with the fall class focusing on print documents and the spring class on digital documents. For the final project in spring semesters, student teams built a website; in fall semesters, student teams designed a collection of print documents, such as a brochure, a collection of stickers, a flier, and an invitation. It was one of the only technical communication courses with a prerequisite: the technologies class in which students developed familiarity with design software, among other tools. This prerequisite enabled a course emphasis on applying design knowledge by producing complete documents: e.g., not just mocking up a website layout but building a working website and publishing it online. In the early course design, clients were selected by students: e.g., a student club or a small business owned by a family member.

During the 2013–2014 academic year, the TPC faculty at USU re-envisioned our program to incorporate a focus on social justice throughout the curriculum. We developed new courses and redesigned existing courses to reflect this focus. To explore student perspectives of social justice and its relevance to their own professional practice, Walton et al. (2017) conducted an IRB-approved pedagogical study. The Fall 2014 and Spring 2015 sections of Document Design and Graphics were two of the courses analyzed in the pedagogical study and iteratively redesigned to reflect the program's new focus on social justice. Refer to Walton et al. (2017) for a Programmatic Showcase article describing the program transition in detail.

Core of Design

In the decade since the programmatic redesign, the Document Design and Graphics course has undergone several iterations. Other courses in our undergraduate curriculum now involve designing, creating, or modifying digital documents such as websites, videos, podcasts, and social media posts. So, the design course now typically emphasizes designing for print to offer students the opportunity to develop a print-design experience in the curriculum. It is almost always taught as a service-learning course in partnership with a local nonprofit organization or on-campus unit, with the final project shaped by client needs. The design thinking process informs the series of iterative assignments that build toward the materials students hand off to clients at the end of the semester. And the course now has a new title: Visual Communication Design. In this course, a consistent goal is to normalize inclusive design practices among early-career technical communicators: i.e., undergraduate students. Two ways we pursue this goal are to (1) thread considerations of justice throughout the course and (2) demonstrate the relevance of inclusive design to student goals.

Rather than treating inclusion as something "extra" or distinct from other considerations of design, we thread inclusion throughout the course by raising specific inclusion issues that are applicable to the stage of design thinking or the particular design topic we're addressing that week. Take, for example, image selection. Lessons on image selection include topics such as how graphics suggest meaning, the strengths and weaknesses of various file types (e.g., when to use jpeg versus tif), and relevant effects on users such as the picture superiority effect: how images support users' recognition and recall of information, particularly when images and text reinforce the same message (Lidwell et al., 2010). We also discuss relevant dangers: how images can reinforce stereotypes, tokenize marginalized communities, and lack relevant representation. In so doing, such images can alienate users or inaccurately signal irrelevance to people who may otherwise use documents. Class discussions, then, prompt students to grapple with the complexities of image selection, recognizing that some strengths of images—such as supporting users' ability to remember content—can amplify the damage of problematic imagery. This framing—incorporating specific considerations of inclusion with their respective design topic—also helps to normalize inclusion as part of the design process because inclusion, in this case, becomes a key consideration of audience awareness.

Similarly, lessons on color address symbolisms of color, how the human visual system perceives color, how color is reproduced by various print technologies versus on screen, and the vocabulary for different types of color schemes and modes. Lessons on color also address color vision deficiency and the importance of dual coding: i.e., visually distinguishing information in more than one way, such as using both shape and color, to make information accessible to a wider range of users. In class, students get hands-on practice with tools such as the free online resource Adobe Color. As instructors, we point out that this industry tool uses vocabulary they just learned (such as "split complementary" color scheme), and we demonstrate features of the tool such as translating across color modes and checking the accessibility of figure-ground contrast for various sizes of text. Student teams practice using the tool in class, submitting a potential color scheme for an upcoming assignment, along with the proper term for the type of color scheme (e.g., analogous), the value for each color in a mode appropriate for the type of document (e.g., the CMYK value for each color if it is a print document), and the contrast ratio for background and foreground colors to confirm that it meets accessibility standards. In this way, inclusion is normalized alongside other design knowledge such as vocabulary and tool use.

The color activity example also demonstrates the second strategy we use to normalize inclusive design: Connecting it to student goals. Many undergraduate students have pragmatic, career-focused goals. They want to learn about design to add to their practitioner skillset, and some students are skeptical that inclusion is truly relevant to industry practice or figures into the work of practitioners. So, we seek out readings by and for industry professionals—particularly those involving inclusive design: e.g., blog articles such as "6 Design Failures that Could Have Been Avoided with Inclusive UX" and videos like Google's "Unconscious Bias at Work." We have found that when practitioners make the case for inclusive design, that case can seem more credible to students. Therefore, we engage strategies such as selecting readings targeted to practitioners and inviting guest speakers from industry to reinforce the message that best practice is inclusive practice. And, to be fair, many undergraduate students *are* committed to inclusive practice because it aligns with their own values. But these students may not recognize *how* inclusion can inform and improve design practice: i.e., how, exactly, to "do" inclusive design. Both of these goals–learning how to "do" inclusive design and getting practical design experience—are supported by applied learning in which they can design real materials for actual clients: enacting the second of Tham's (2021) pedagogical heuristics—designing actionable solutions—by way of his first heuristic—engaging in iterative processes.

Design Activities and Learning Exercises

One assignment that supports several learning objectives, including normalizing inclusive design, is the card deck assignment. It begins with an online reading by Airbnb Design. In 2016, Airbnb publicly apologized for its slow response to widespread customer reports of racism, in which Airbnb hosts canceled or declined reservations based on guests' skin color. CEO Brian Chesky acknowledged, "There were lots of things we didn't think about when we, as three white guys, designed the platform" (Solomon, 2016). In addition to developing an anti-discrimination policy, the Airbnb design team partnered with a journalism startup to develop a tool that could help them recognize biases and related problems earlier in the design process. This tool, called "Another Lens," sets out three guiding principles for design: Balance your bias, Consider the opposite, and Embrace a growth mindset. Each principle has five to six associated questions, such as, "Who's someone I'm nervous to talk to about this [design]?" Each question is accompanied by a brief paragraph explaining the reasoning behind the question and the kinds of biases the question aims to mitigate. The reading is brief; the language is understandable; and the tool is actionable. We often pair this reading with one or two additional readings on bias in design, using these readings to frame and inform discussions of both the empathy stage and the ideate stage of design thinking.

For the assignment, each student team designs a card deck based on the Another Lens tool:

- For this assignment, your team will design a deck of cards that you can use as a design resource moving forward. Using content from the Another Lens reading, you'll produce one card for each of the 16 questions, with the three guiding principles serving as "suits" for the card deck.
- Each card will be double-sided. One side should convey the question and a brief explanation of how and why that question can inform inclusive design. (This content should be quoted or paraphrased from the reading.) On the other side, provide one or two brief examples of a design situation in which such a question could be useful. (This content should be developed by you.)

It can be challenging for students to imagine a design scenario targeted to each question, and this challenge leads to some rich class discussions in which students help each other to consider how each reflection question could lead to a more inclusive design outcome, reveal potential biases, or otherwise improve designs. Each team prints one set of cards for each member, which offers two benefits. First, students get experience navigating the print-production process early in the semester. Inevitably, complications arise: e.g., communication challenges, timeline snafus, or surprises related to color reproduction or paper transparency. These challenges are never welcome, but it can be less stressful for students to encounter them for the first time on a lower-stakes assignment before the end-of-semester crunch. Second, student teams get their own card deck to use in their design process moving forward (discussed in the Design Activities and Learning Exercises for Professionally Printed Materials below).

Customizing Options

As Turner and Rose (2022) found in their study instructors approach UX pedagogy in different ways, we build on the Visual Communication Design course described above and ask instructors to customize the course in various ways that suit their teaching preferences. To assist readers with envisioning such customizations, we provide two examples below to show how Alexander and Walton customized the course at our respective institutions.

Partnerships with Organizations that Promote Inclusion

Knowing that there have been opportunities for learning TPC service-learning courses through "summer institutes (short courses), by NCTE and

CCCC, by ATTW, STC, and IEEE, by publications, and by preparation in graduate programs" (Pickett, 1997, p. 291), we were inspired to create a service-learning course that promoted inclusive excellence for community partners. With this goal, Alexander and Walton have partnered with various campus and community partners, seeking out partner organizations in need of design specialists and creating chances for reciprocity. In other words, not only do our community partners benefit from student design labor, but students get the chance to gain experience with creating designs that will be used by professional entities.

It's important to note that the term "community partner" encompasses a wide range of stakeholders within and around a university. For example, Alexander spent two+ years developing relationships with the Inclusion Center at USU, one year with the Multicultural Student Life Center at the University of Tennessee, Knoxville (UTK), as well as non-university partners with small businesses and non-profit organizations. The relationships developed provide Alexander with unique opportunities for students to work with an actual client and the documents they need assistance with. Similarly, Walton has partnered with the USU's Writing Center and the Department of English where documents are always needed for different purposes and where clients have experience working with students.

Key Considerations of This Version

Alexander developed relationships with community partners who have gone through several different iterations of marketing and have struggled to get input from graphic designers. By working with said partners, students learn how to best create different advertisements and effective marketing techniques that appeal to specific audiences. With this problem in mind, Alexander created the following question that is addressed throughout his collaborations: *How can this course assist our community partner with increasing their level of visibility so that minoritized stakeholders can know that the organization/establishment exists and plays an important role in advocating and supporting all community members?*

Each collaboration exists as a means to bring ownership to the learning process and enable students to experience transferable skills most needed in the community and industry. Alexander introduces students to inclusive practices to help effectively design visual documents that appeal to divergent communities, allowing them to use different terministic screens[1] that center the marginalized. He is explicit about inclusive learning objectives, explaining in the syllabus that students will be

(a) exposed to new realities that push against their cultural norms, (b) identify (in)effective design practices/choices to help them select the most appropriate options based on their community partner's needs, (c) learn how to coexist and help other non-minoritized stakeholders co-exist with underrepresented communities, (d) play a part in enhancing inclusive excellence within their community, and (e) learn how people are represented via marketing materials while understanding *how* that representation influences design decisions.

To support students in this process, Alexander requires them to collaborate in their community partners' workspace multiple times throughout the semester to learn more about the space they'll be designing for while interacting with the target audiences as a way to design with them in mind. In other words, students are taught to empathize by acquiring insights into their community partners' world, creating chances for them to develop a growth mindset. This move aligns with Tham's (2021) third pedagogical heuristic of design thinking which prioritizes open spaces for collaboration. Students have various opportunities to apply their skills within complex, real-world contexts by working in these spaces to develop a set of print and digital documents while pitching the redesign of various documents, including window clings. For example, in 2021, he had students create window clings for the Inclusion Center at USU as a marketing/recruiting tool to attract minoritized students to the different clubs that promoted and celebrated their identities. In 2023, he had students do the same for the Multicultural Student Life Center at UTK, but on a larger scale. Students designed window clings to be featured in the main lobby of the center, which is situated in a major crosswalk, allowing the center to promote and attract incoming students to "B.E. S.E.E.N.[3]"

Relevant Design Activities

To promote inclusive design, Alexander created opportunities for students to explore culturally sensitive topics about designs that (un)intentionally excluded communities often found in the margins. Such activities help new designers understand "factors such as race, gender, sexuality, class, age, and ability—and intersections thereof—all impact the way people experience" designs/products (Iyamah, 2023, para. 1). By explicitly accounting for considerations of intersectionality in the design thinking process,

Alexander implements a version of what Lane (2021) advocates in her interstitial design framework. Specifically, Alexander scaffolds student engagement with a series of *DOGO Inclusive Design* activities, created by Norwegian scholars. Each activity takes an industry approach to help students add inclusive design to their design process with eight steps that are broken into four sections (Design and Architecture Norway, 2018). These activities teach students how to engage in more inclusive research, supporting the notions of the card deck assignment discussed above. We provide an overview of the four sections below and invite readers to consider these activities when designing for social justice:

Section 1: Explore

- Understanding Context
- Designing Research

Section 2: Focus

- Discovering Needs
- Mapping Insights

Section 3: Develop

- Translating Briefs
- Scenario Building

Section 4: Deliver

- User Feedback
- Resource Building

Each section takes a people-centered approach while explaining how students can incorporate each step into the various stages of design (Design and Architecture Norway, 2018). In other words, the eight steps (two per section) provide students with an overview of practical techniques for bringing inclusive design thinking into their process. The activity takes place over two-course periods:

> In the [first two sections], these activities include methods for setting up your research, obtaining insights from users[,] and then using these insights as inspiration for ideas. In the later [sections], as [your] project[s] progresses through [the] development cycles, the activities help to maintain a people-centered perspective during decision-making and evaluation.
>
> *(Design and Architecture Norway, 2018, para. 4)*

By the end of these activities, students are better equipped to have effective and transparent conversations with their community partners, which enhances their chances of taking their client's ideas and making them equitable across multiple avenues.

Professionally Printed Materials

Another version of the course design involves securing a small amount of funding to enable student teams to produce professionally printed final materials. Below we discuss some considerations relevant to this version of the course, followed by a couple of assignments designed to prompt and reward iteration.

Key Considerations of This Version

Key to this version of the course is securing a small amount of funding that enables student teams to professionally produce their final materials. Even a small print budget can benefit both students and client organizations. For students, it increases equity by ensuring that every team member gets the experience of managing a budget and working with a professional print shop–regardless of whether students themselves can afford professional printing. For client organizations, it offers valuable reciprocity. A common challenge of service-learning partnerships is ensuring the community partner truly benefits, especially given that service-learning partnerships can be time-consuming for community organizations (Bourelle, 2014). With this course design, not only do client organizations receive free design expertise, but they also receive professionally printed copies of the materials. In this way, even a small amount of funding can help to facilitate the radical collaboration encouraged by Tham et al. (2022) by making service-learning partnerships feasible for a wider range of community partners. This reciprocity can make the partnership worthwhile for small community organizations that lack design expertise and have little to no print budget. And if a client organization needs more copies of a particular item than the small class budget will cover (e.g., 300 invitations), clients still benefit: they know exactly what the final product looks like, what it costs, and where to get additional copies printed with just a phone call.

In addition to client reciprocity, it's also important to set students up for success by ensuring that course learning objectives, assignment descriptions, client needs, and production budgets align. To identify the amount of necessary funding, Alexander and Walton met with their respective print shop directors before the beginning of the semester to describe relevant assignments and consult them regarding an appropriate budget. Walton then used the course cap and an estimated team size of 3–5 students to calculate the total amount of necessary funding, which was approximately $750 total for the class.

In seeking potential sources of funding, Walton found that several university entities may offer small internal grants to support student learning. For example, inclusion centers may award funding for classes with learning objectives relevant to the center's mission. The library, instructional design center, center for community engagement, or office of teaching excellence also may offer small grants to support experiential learning. Alternatively, discretionary funds may be available at the department or college level, allowing instructors to request funding without a formal proposal or application process. Alexander has employed a different approach by meeting with the client organization before the semester to explore their needs and securing a $500 budget for each student team funded directly by the client.

Managing their small budget requires students to develop coordination strategies for different stakeholder groups. Students coordinate with clients and users regarding needs, with the instructor regarding assignment requirements, with the print shop regarding capacities and cost, and with each other regarding the workload. To support coordination, students develop a team charter (Wolfe, 2010) with guidelines for team and client communication, iterative deadlines, and procedures for decision-making and conflict resolution. When the clients are university entities, Walton requires that students submit their designs for review by the university copyright and trademark team so the materials are approved for official use. This increases the value of the assignments as portfolio content for students, who can discuss their engagement in the design thinking process from initial client meetings, through iterations of design and testing, to organizational review and approval, to professional printing. And because their materials are officially approved for use by the trademark and copyright team, students are sowing a legacy of inclusive university materials. However, layering in these real-world considerations—budget constraints and organizational approval—makes it challenging to fit projects within the semester timeframe. Careful scaffolding is essential to ensure that the design thinking process—especially empathy and ideation—isn't clipped in the interest of time. Below we describe how we build this scaffolding into the course design.

Relevant Design Assignments

We encourage iteration through assignment design, highlighting the fourth step in Tham's (2021) pedagogical heuristics—where design thinking encourages students to focus on human-centered design. For example, the card deck assignment sets up student teams to engage in reflection exercises that prompt and inform iteration of their design to improve inclusivity. The subsequent assignment requires them to submit photos of the team engaging with the cards, along with a brief write-up about which cards they used and how their design was affected[4]. If the instructor knows they will miss a particular class (for example, to attend a conference), they can schedule this assignment as a self-directed activity that students can conduct during regular class time.

In addition to the card deck, two other assignments prompt iterative design, albeit in different ways: the iteration assignment and the revision plan. These are described below.

The iteration assignment requires students to compile documentation of their iterative work throughout the semester. These materials range widely–from quick sketches scratched out during team meetings to client emails summarizing their discussions to design drafts both hand-drawn and digital. In other words, this assignment requires students to hang onto materials they produce at each stage of the design thinking process. Walton introduces the assignment early in the semester and periodically reminds students to keep track of their documentation. She explains two of the benefits this assignment is designed to offer students. First, evidence of iterative design (especially sketches and early prototypes) can be valuable content for students' professional portfolios. After all, it is the design *process* rather than the design *artifact* that is more likely to be transferable across professional contexts. Evidence of the design process, then, is valuable portfolio content illustrating the professional competencies students can apply in the workplace, but this value may not be initially apparent to students, who may otherwise discard these materials as they move closer to a final design. Second, this assignment gives students credit for work they should be doing anyway. If student teams engage in the design thinking process and hang onto the detritus they develop during that engagement, this assignment basically provides "free points" toward their grade. In this way, the assignment *prompts* students to engage in the design thinking process and *rewards* them for doing so.

Student teams also submit drafts at two points during the semester. Earlier in the semester, teams sign up for a meeting with the instructor to share their first draft and discuss their next steps toward a final design. Later in the semester, student teams engage in peer reviews of more polished drafts. The key to these draft assignments is the two-part revision plan, which is due within 24 hours after the drafts. Part one of the revision plan distills feedback from the instructor or peers; part two briefly describes subsequent revisions the team will make based on the feedback and additional information from clients or users they will seek to inform their next steps. Thus, the revision plan prompts teams to document the feedback they found most useful or thought-provoking and to overview how that feedback will inform the design process moving forward. This assignment is valuable for ensuring that draft reviews do, in fact, prompt design iterations.

Conclusion & Future Versions

Throughout this chapter, we have discussed ways we aim to normalize inclusive design practices in variations of a design-thinking course we teach at our respective institutions. In addition to providing examples of our activities and assignments, we call instructors to reflect upon how they can adapt such approaches to their own classes and contexts, as well as to identify relevant considerations that are under-addressed throughout their curriculum.

In reflecting on our future iterations of this course design, Anabire will develop opportunities for students to consider localization and globalization strategies in the design thinking process. Such opportunities will answer calls such as that of Flammia (2011), who argues for TPC instructors to train students for future work in cross-cultural and multicultural environments. As TPC programs become more diverse in terms of curriculum, student populations, and alumni careers, students are expressing interest in working for international organizations, particularly nonprofits (Jones et al., 2014). Collaborating with international partners in service-learning courses can help students connect and engage with global communities (Baniya et al., 2021), which can support students in building relevant professional networks. Additional potential benefits include developing cultural sensitivity, practicing intercultural communication, and developing competencies with collaboration technologies (Flammia, 2011). Thus, Anabire plans to pursue partnerships outside the US to allow for service-learning courses that address localization and globalization issues in more depth. Such partnerships can provide a specific context for cross-cultural considerations of design and extend our current efforts to normalize inclusive design into international contexts.

Notes

1 Card sorting is a "UX research method in which study participants group individual labels written on notecards according to criteria that make sense to them" (Sherwin, 2018, para. 2).
2 Terministic screens refer to the different ways people view the world. In other words, "any given terminology is a reflection of reality, by its very nature as a terminology it must be a selection of reality; and to this extent it must function also as a deflection of reality" (Burke, 1965, p. 88).

3 The acronym B.E. S.E.E.N. is an initiative that [name of center] promotes "Belonging within the campus community. We hope to champion all [redacted mascot name] as they Excel both inside and outside of the classroom. We want to ensure that all [redacted mascot name] feel Supported throughout their tenure here on [redacted name of city!] Our office hopes to Empower all [redacted mascot name] through meaningful and intentional Experiences in the form of various programs and initiatives while allowing for our space to serve as the Nucleus for campus inclusion, education, identity, and diversity.

4 The photos are not essential to the design of this small assignment, but Author 3 has found that student teams seem to enjoy taking funny photos of themselves with the cards and sending them to the instructor. This is even true when student teams meet online through software such as Zoom or Teams and take screenshots of themselves holding cards and making faces.

References

Baniya, S., Brein, A., & Call, K. (2021). International service learning in technical communication during a global pandemic. *Programmatic Perspectives*, *12*(2), 26–58.

Bay, J., Johnson-Sheehan, R., & Cook, D. (2018). Design thinking via experiential learning: Thinking like an entrepreneur in technical communication courses. *Programmatic Perspectives*, *10*(1), 172–200.

Bourelle, T. (2014). Adapting service-learning into the online technical communication classroom: A framework and model. *Technical Communication Quarterly*, *23*(4), 247–264. https://doi.org/10.1080/10572252.2014.941782

Burke, K. (1965). Terministic screens. In *Proceedings of the American Catholic Philosophical Association* (Vol. 39, pp. 87–102). American Catholic Philosophical Association.

Cook, K. C. (2002). Layered literacies: A theoretical frame for technical communication pedagogy. *Technical Communication Quarterly*, *11*(1), 5–29. https://doi.org/10.1207/s15427625tcq1101_1

Design and Architecture Norway. (2018, January 18) *Innovating with people - inclusive design and architecture*. Design and Architecture Norway. https://doga.no/en/tools/inclusive-design/tools-and-methods/adding-inclusive-design-to-your-design-proccess/

Flammia, M. (2011). Using service-learning and global virtual team projects to integrate sustainability into the Technical Communication Curriculum. In the *Proceedings of 2011 IEEE International Professional Communication Conference*. https://doi.org/10.1109/ipcc.2011.6087192

Hutter, L., & Lawrence, H. M. (2018). Promoting inclusive and accessible design in usability testing. *Communication Design Quarterly*, *6*(2), 21–30. https://doi.org/10.1145/3282665.3282668

Iyamah, J. (2023, June 7). 6 Design failures that could have been avoided with inclusive UX research. https://www.userinterviews.com/blog/design-failure-examples-caused-by-bias-noninclusive-ux-research

Jones, N., Savage, G., & Yu, H. (2014). Tracking our progress: Diversity in technical and professional communication programs. *Programmatic Perspectives*, *6*(1), 132–152.

Lane, L. (2018). Iteration for impact. In the *Proceedings of the 36th ACM International Conference on the Design of Communication*. https://doi.org/10.1145/3233756.3233952

Lane, L. (2021). Plotting an interstitial design process: Design thinking and social design processes as framework for addressing social justice issues in TPC classrooms. In *Equipping technical communicators for social justice work: Theories, methodologies, and pedagogies* (pp. 214–229). Utah State University Press.

Lidwell, W., Holden, K., & Butler, J. (2010). *Universal principles of design, revised and updated: 125 ways to enhance usability, influence perception, increase appeal, make better design decisions, and teach through design.* Rockport Publishing.

Moeggenberg, Z. C., & Walton, R. (2019). How queer theory can inform design thinking pedagogy. In the *Proceedings of the 37th ACM International Conference on the Design of Communication.* https://doi.org/10.1145/3328020.3353924

Morgan, M., Allen, N., Moore, T., Atkinson, D., & Snow, C. (1987). Collaborative writing in the classroom. *The Bulletin of the Association for Business Communication, 50*(3), 20–26.

Pickett, N. A. (1997). The technical communication service course serves. In K. Staples, & C. M. Ornatowski (Eds.), *Foundations for teaching technical communication: Theory, practice, and program design* (pp. 287–98). Bloomsbury Publishing.

Sherwin, K. (2018, March 18). Card sorting: Uncover users' mental models for better information architecture. Nielsen Norman Group. https://www.nngroup.com/articles/card-sorting-definition/

Solomon, B. (2016, July 13). Airbnb confronts racism as it hits 100 million guest arrivals. Forbes. https://www.forbes.com/sites/briansolomon/2016/07/13/airbnb-confronts-racism-as-it-hits-100-million-guest-arrivals/?sh=73aeceab6b76

Stone, E. (2000). Service learning in the introductory technical writing class: A perfect match? *Journal of Technical Writing and Communication, 30*(4), 385–398. https://doi.org/10.2190/9ed8-hek6-pddl-4gqb

Tham, J. (2021). Engaging design thinking and making in technical and professional communication pedagogy. *Technical Communication Quarterly, 30*(4), 392–409. https://doi.org/10.1080/10572252.2020.1804619

Tham, J., Howard, T., & Verhulsdonck, G. (2022). Extending design thinking, content strategy, and artificial intelligence into technical communication and user experience design programs: Further pedagogical implications. *Journal of Technical Writing and Communication, 52*(4), 428–459. https://doi.org/10.1177/00472816211072533

Turner, H. N., & Rose, E. J. (2022). What do we teach when we say we teach UX? A study of the practices of TPC instructors. *Programmatic Perspectives, 13*(1), 61–102.

Walton, R., Colton, J. S., Wheatley-Boxx, R. K., & Gurko, K. (2017). Social justice across the curriculum: Research-based course design. *Programmatic Perspectives, 8*(2), 119.

Wolfe, J. (2010). *Team writing.* Bedford/St. Martin's.

EDITORS' OUTRO

Opportunities and Challenges of Multimodal Community Engagement for Social Justice

Overall Lessons Learned

This collection is a culmination of wisdom and experience by scholars, practitioners, and community participants who have envisioned, designed, and executed multimodal projects aimed at advancing social justice through community engagement. By way of closing, we spotlight here three high-level observations gleaned from the chapters in this book.

First, the definitions of community-engaged scholarship and their associated frameworks are contextual and should be adapted to local needs. Whether in regional, national, or inter/transnational settings, the design of community engagement should be fostered and maintained by tending to local conditions. From Knight's (Chapter 1) community-led design framework to Phillips's (Chapter 2) environmental justice advocacy strategies, from Carlson's (Chapter 3) place-based community research methods to Koepke's (Chapter 14) reproductive justice digital storytelling, community engagement efforts are situated within specific hopes or requirements of the local partners and community members. Their recommendations can be adopted by multimodal projects elsewhere but must be evaluated for social factors such as economic status, access to education and literacy sponsors, collective or community values, environmental conditions, political morale, and other visible or implicit tensions. When defining their parameters and methodologies for community engagement, project leaders and designers should also pay attention to technical or technological issues, including mobility and transportation systems, material resources and availability (e.g., funding, product supplies), place and spatial orientation (e.g., geography, location, city bounds, neighborhoods), digital and computing infrastructure (e.g., tools, features, support, accessibility), and other functional matters of concern. All these factors contribute to the viability and potential for success in community-engaged projects.

Second, ethical values and principles should drive the design and development of community-engaged projects. As demonstrated by contributors like Azrul-Carmichael et al. (Chapter 7), Holmes and Colton (Chapter 5), Baniya et al. (Chapter 6), Cecil-Lemkin (Chapter 16), Nish (Chapter 9), and Gonzales et al. (Chapter 10), community engagement offers an opportunity to not only serve local needs but enact social justice via advocacy for ethical practices in identity expression, citizenship, accessibility and disability justice, and linguistic representation. Project attributes and engagement mindsets must be founded on the basis of ethics, educating and enabling equitable and just treatment of all in the community. While such enactment of justice might be challenging for some projects and communities, scholars should exercise their power, privilege, and positionality to assume responsibility in engaging the difficult interactions that may emerge from community partnerships. Ethics and ethical practices must be guiding principles of community-engaged activities, not an optional feature.

Third, we consider pedagogy to be the primary interface between students and community partners. Scholar-teachers are well-positioned to facilitate positive and meaningful exchanges among students and community members through project activities that enable productive interactions. These interactions may not happen naturally nor neutrally. As justice-oriented topics remain politically sensitive, teachers should be attentive to student participation—to protect them from undesirable or targeted responses—as well as community (re)actions throughout the project period, and identify pedagogical moments where either or both parties can learn about social justice. Teaching frameworks recommended by the contributors such as Byrd (Chapter 13), Alexander et al. (Chapter 17), Hill et al. (Chapter 8), Grant (Chapter 15), and Moon (Chapter 4), including counterstory, inclusive design, care ethics, and design thinking, can be helpful for designing sound pedagogies that support the objectives of community-engaged projects. It goes without saying, the outcomes of community-engaged collaboration, and its corresponding pedagogy should benefit learning and service in a mutual fashion.

Furthermore, critical attention should be paid to community-engaged design projects in global and transnational contexts, which involves recognizing and addressing the diverse cultural, socio-political, and economic factors that influence such projects across different regions and populations. From how Friday et al.'s descriptions of Wikitongues's innovative approach toward the global revitalization of endangered languages (Chapter 11) to Yang and Wang's (Chapter 12) discussions of Chinese social media platforms' impacts on grassroot activism, it is essential to understand how global power dynamics, historical contexts, and local realities shape the communities involved in these projects. Scholars and practitioners should engage in thoughtful analysis of how their design practices and solutions can play an active role in navigating and addressing existing inequalities. By being mindful of these factors, researchers can ensure that their work is not only culturally sensitive and contextually relevant but also capable of fostering transnational collaboration and empowerment. This approach encourages the development of more inclusive, equitable, and impactful design strategies that address both local and global social justice issues, facilitating meaningful engagement and positive change across diverse communities on a global level.

Future Opportunities and Challenges in Designing for Social Justice

Designing for social justice is indeed a multifaceted endeavor—one that intertwines challenges, hurdles, and meaningful impact. One of the challenges lies in overcoming the deficit mindset often visible in community-engaged collaborations with marginalized populations. In our earlier research on technical and professional communication (TPC) instructors' feelings toward community-engaged projects (Jiang & Tham, 2023, 2024), we have observed that racialized emotions, such as pity and disgust, persist in professional documentation and student attitudes toward marginalized communities. These documents often convey negative emotions toward communities of color. More specifically, our analysis of interviews with TPC instructors reveals that professional documentation perpetuates a deficit mindset, portraying communities of color as inherently lacking. Building on previous discussions of the deficit mindset (Espino & Lee, 2011; Valencia, 2010), our study demonstrates how this mindset is effectively shaped and reinforced through professional documentation and training in community-engaged projects. Feelings of pity and sympathy often stem from the deficit mindset, which is rooted in the notion of white superiority over people of color and the belief that it is the responsibility of white individuals to help those who are less privileged. Beneath seemingly benign emotions, including pity, lies a deeply ingrained sense of disgust directed at racialized minority individuals (Ahmed, 2004; Matias & Zembylas, 2014), which reinforces existing power imbalances and sustains oppressive systems.

Navigating these challenges requires critical emotional awareness. As TPC scholars and practitioners, we grapple with these ongoing questions for discussions: How can we leverage multimodal theories, such as design thinking, to amplify marginalized voices? How can we dismantle oppressive ideologies and narratives through multimodal design? How might our design artifacts and user interfaces foster inclusivity and empathy? Design thinking can be a powerful tool for helping teachers and students empathize with marginalized communities by fostering a deeper understanding of their experiences and needs. For instance, instructors may invite students to conduct interviews with community members, listen to their stories, and understand their perspectives. This human-centered approach involves immersing designers in the perspectives of those directly affected by social issues, encouraging active listening and engagement. Teachers and students may also involve community members in the design process and create innovative design solutions that resonate with their stories. Through processes such as empathy mapping, user interviews, and user research, teachers and students can uncover the lived histories and experiences of marginalized individuals.

Ultimately, we recommend that scholars and practitioners include multimodal theory and practice in a variety of contexts to support writing and designing for social justice. By integrating diverse modes of communication—such as text, visuals, audio, and interactive elements—into their work, we can create more engaging, accessible, and impactful community-engaged projects. Scholars, instructors, and practitioners can encourage students to create counter-storytelling projects that combine text, visuals, audio, and video to amplify marginalized voices. For example, digital storytelling (Jiang, 2024a, 2024b; Matias & Grosland, 2016; Rolón-Dow, 2011) leverages students' existing technological skills to create innovative expressions in the form of videos and multimedia stories. Our study of TPC instructors' feelings toward designing for social justice (Jiang & Tham, 2023, 2024; Tham & Jiang, 2022, 2024) demonstrates how counter-stories can effectively challenge the emotional dynamics of whiteness concerning communities of color. These projects enable students to engage empathetically with the viewpoints of these communities, including racialized marginalized and immigrant groups. Instructors might encourage students to interview people of color about their experiences, develop stories that reflect these perspectives, and consider how these counter-stories help dismantle racialized emotions. Presenting counter-stories to an external audience can also help students understand the roots of discrimination and deepen their engagement with challenging topics such as racial oppression.

References

Ahmed, S. (2004). *The cultural politics of emotion*. Edinburgh University Press.

Espino, M. M., & Lee, J. J. (2011). Understanding resistance: Reflections on race and privilege through service-learning. *Equity & Excellence in Education*, *44*(2), 136–152.

Jiang, J. (2024a). Emotional landscape of translingualism: Multilingual international students navigating shame through translingual digital stories. *TESOL Quarterly*. Early view. https://onlinelibrary.wiley.com/doi/10.1002/tesq.3341

Jiang, J. (2024b). "Emotions are what will draw people in": A study of critical affective literacy through digital storytelling. *Journal of Adolescent & Adult Literacy*, *67*(4), 253–263. https://doi.org/10.1002/jaal.1322

Jiang, J., & Tham, J. (2023). Rethinking multimodal community-engaged pedagogy through posthumanist theory. *Teaching in Higher Education*. 1–17. https://doi.org/10.1080/13562517.2023.2253758

Jiang, J., & Tham, J. (2024). Race, affect, and marginalized communities: Navigating racialized emotions in community-engaged pedagogy. *Critical Studies in Education*, 1–19. https://doi.org/10.1080/17508487.2024.2343284

Matias, C. E., & Grosland, T. J. (2016). Digital storytelling as racial justice. *Journal of Teacher Education*, *67*(2), 152–164. https://doi.org/10.1177/002248711 5624493

Matias, C. E., & Zembylas, M. (2014). 'When saying you care is not really caring': Emotions of disgust, whiteness ideology, and teacher education. *Critical Studies in Education*, *55*(3), 319–337. http://dx.doi.org/10.1080/17508487.2014. 922489

Rolón-Dow, R. (2011). Race (ing) stories: Digital storytelling as a tool for critical race scholarship. *Race Ethnicity and Education*, *14*(2), 159–173.

Tham, J., & Jiang, J. (2022). Examining multimodal community-engaged projects for technical and professional communication: Motivation, design, technology, and impact. *Journal of Technical Writing and Communication*, *53*(2), 128–159. https://doi.org/10.1177/00472816221115141

Tham, J., & Jiang, J. (2024). Understanding writing instructors' feelings toward the affordances of multimodal social advocacy projects: Implications for service-learning pedagogies. *College Composition and Communication*, *76*(1), 4–34.

Valencia, R. R. (2010). *Dismantling contemporary deficit thinking: Educational thought and practice*. Routledge.

GLOSSARY

A–Z Concepts for Community Engagement

Accessibility is the intentional design and implementation of initiatives to ensure that all people regardless of their abilities can fully participate in and benefit from community-engaged projects. Accessibility involves creating inclusive environments that accommodate diverse needs, removing barriers to participation, and providing necessary resources and support.

Benefits refers to the positive outcomes and advantages gained by the participants involved in community-engaged projects. These benefits can include enhanced skills, knowledge, and personal growth for participants; improved services, resources, and overall well-being for the community; strengthened relationships and collaboration between institutions and community members; and the promotion of social justice, equity, and sustainable development.

Community is a group of people who share common geographic locations, interests, values, or identities. This group can include residents, local organizations, schools, businesses, and other stakeholders who collaborate to address shared challenges, leverage local assets, and achieve mutual goals.

Design includes the process of creating and planning solutions, systems, or interventions that address the needs, aspirations, and challenges of a community. It involves collaborative and iterative approaches where community members actively participate in shaping and informing the decision-making process.

Ethics signals the principles and standards that guide the conduct of individuals and organizations involved in community-engaged projects. It encompasses the commitment to respect, fairness, and integrity in interactions with community members, ensuring that their rights, dignity, and voices are upheld.

Facilitation involves the efforts to ensure effective communication, participation, and problem-solving among all participants of community-engaged projects. Facilitators help to structure discussions, manage group dynamics, and address conflicts or barriers that may arise, with the aim of achieving productive outcomes and fostering a sense of shared ownership and commitment.

Grassroots refers to initiatives and movements that originate from and are driven by local community members rather than external authorities or institutions. This approach emphasizes the involvement and leadership of individuals directly affected by the issues at hand, ensuring that the solutions and strategies developed are relevant, responsive, and tailored to the community's unique needs and context.

Humanistic projects emphasize the importance of understanding and addressing the human dimensions of community issues. This perspective prioritizes the values, experiences, and well-being of individuals and communities, focusing on empathy, respect, and the holistic aspects of human life.

Inclusivity is the practice of ensuring that all participants, regardless of their background, identity, or circumstances, have equitable access to and meaningful involvement in the project. Inclusivity aims to foster a sense of belonging, ensure that all participants can contribute and benefit, and promote equitable outcomes by considering and addressing the needs and contributions of all community members.

Justice means striving to create equitable partnerships, supporting the rights and dignity of all participants, and focusing on creating positive social change that addresses the root causes of disparities. It also includes advocating for policies and practices that promote inclusivity and equitable outcomes for all members of the community.

Kindness involves fostering a supportive and respectful environment, where participants are treated with compassion and understanding. It means prioritizing positive relationships, listening actively, and addressing the needs and concerns of others with sensitivity and respect.

Literacy typically refers to the ability to access, understand, and utilize information and resources effectively within a specific context. In community-engaged projects, literacy involves equipping community members with the knowledge and skills needed to participate meaningfully in the project.

Multimodality is the use of multiple modes of communication and expression to enhance understanding, participation, and collaboration. It is a concept that recognizes how individuals may engage with and process information through various channels, such as text, visuals, audio, and interactive media. By employing multiple modes, community-engaged projects can achieve different rhetorical effects.

New media are digital platforms and technologies that facilitate interactive and dynamic forms of communication and collaboration. This includes social media, blogs, podcasts, videos, and other online tools that enable real-time engagement, information sharing, and community interaction. New media can be used to enhance outreach, foster dialogue, gather feedback, and document project progress.

Outcomes are the measurable and observable results or impacts that arise from the implementation of community-led designs or projects. These outcomes encompass the changes or benefits experienced by the community and participants as a result of the project. Evaluating outcomes involves assessing both short-term and long-term effects to determine the effectiveness and success of the engagement efforts and to inform future projects.

Publics highlight the importance of recognizing the heterogeneity within and among communities, referring to the diverse groups of individuals and communities that are the intended or actual audience for engagement efforts. Understanding and addressing the needs, perspectives, and interests of different publics is crucial for ensuring that the engagement activities are relevant, inclusive, and effective.

Queer refers to an inclusive and fluid understanding of gender identities and sexual orientations that challenges traditional binary categories and norms. Integrating a queer perspective involves recognizing and valuing the diverse identities and experiences of LGBTQ+ individuals, addressing specific needs and issues faced by queer communities, and fostering environments of acceptance and inclusivity. It also means actively challenging heteronormative assumptions and practices.

Reflection is the process of thoughtfully examining and analyzing one's experiences, actions, and outcomes related to the community project. It involves critically considering what was learned, what worked well, what challenges were encountered, and how these insights can inform future practice.

Social aspects of community-engaged projects include the interactions, relationships, and structures within a community. It focuses on the ways in which people connect, collaborate, and engage with one another, as well as the social dynamics, networks, and cultural practices that influence community life.

Transformation refers to significant and meaningful changes in individuals, communities, or systems that result from the engagement process. Transformation denotes the process through which projects lead to enduring impacts. It emphasizes the goal of creating positive, sustainable change that goes beyond immediate outcomes and fosters long-term development and empowerment within the community.

Users in the context of community engagement refers to the individuals or groups who interact with, benefit from or are directly impacted by a project. Users are the primary audience whose needs, preferences, and feedback are integral to the project's development and implementation. Understanding and addressing the needs and perspectives of users is crucial for ensuring that the engagement efforts are relevant, effective, and responsive to the community's context.

Virtue is the adherence to principles that guide positive and effective interactions with community members. Virtue involves demonstrating a commitment to ethical practices, ensuring that actions and decisions are guided by a sense of responsibility and care for the well-being of all participants.

Workshops are a participatory method to educate, train, and collaborate with community members. Workshops can include activities such as presentations, group exercises, brainstorming sessions, and skill-building tasks, and are intended to foster capacity-building, knowledge-sharing, and collaborative problem-solving within the community.

X-factor community events are activities that showcase unique talents, skills, or attributes of community members, fostering a sense of pride and connection. These events often feature creative approaches, unconventional formats, or compelling elements to generate enthusiasm, engagement, and meaningful connections among participants.

Yielding means acknowledging and integrating input from participants, adjusting approaches as needed, and prioritizing the needs and preferences of the community. It involves being flexible and responsive in project planning and implementation to ensure that the engagement efforts are effective and respectful of diverse viewpoints.

Zest refers to the enthusiasm, energy, and passionate commitment brought to the planning, execution, and evaluation of community-engaged projects. It involves approaching projects with a vibrant and proactive attitude, inspiring and motivating participants to engage and contribute actively. Zest can enhance the overall impact of the project by fostering a positive and delightful atmosphere.

APPENDIX
Sample Course Syllabus

Empathy in Action: Community-engaged Design for Social Justice

Course Description

This is a 16-week community-engaged course that introduces students to design thinking and multimodal design through hands-on projects aimed at promoting social justice. In this course, students will learn the processes of brainstorming, designing, and revising by working on four major assignments: an advocacy poster, an interview podcast, a promotional video, and a website portfolio. Through peer review activities, the class will work together to critique and improve each other's work. As the course progresses, students will navigate their way around Adobe Photoshop, Audition, Premiere Pro, and Portfolio, learning the ins and outs of photo, audio, and video composition and editing.

Learning Objectives

- *Understand and Apply Design Thinking Principles*: Demonstrate a thorough understanding of design thinking methodologies and apply these principles to community-engaged projects.
- *Promote Empathy through Design*: Use design tools and techniques to create materials that foster empathy and raise awareness about social justice issues.
- *Develop Technical Skills in Design Platforms*: Gain proficiency in platforms such as Adobe Photoshop, Audition, Premiere Pro, and Portfolio for creating and editing visual, audio, and video content.

- *Collaborate Effectively in a Community Context*: Work collaboratively with peers and community partners to design and execute projects that address real-world social justice challenges.
- *Reflect on the Role of Design in Social Change*: Critically reflect on the role of design in social change and the ethical considerations involved in community-engaged design projects.

Course Materials

- This book
- Subscription to Adobe Creative Cloud and/or alternative design platforms
- LinkedIn Learning Tutorials for Adobe Creative Cloud and/or alternative design platforms

Assignments & Activities

Empathetic Audience Analysis: This analysis asks students to engage in the design thinking process of defining the audience's needs, including, among others, the rhetorical purposes of the multimodal design projects, the kind of style needed given the audience and the context, the background information needed for the audience to navigate the design, as well as the potential problems and confusions that audiences may experience. (Recommended program: Microsoft Word or Adobe InDesign).

Advocacy Poster: This project asks students to advocate for a cause through the juxtaposition of images and text. In collaboration with community partners, students could be making a social commentary, raising public awareness, and/or calling for action. Students should consider the community partners' needs and what they would find persuasive or interesting. (Recommended design program: Adobe Photoshop).

Interview Podcast: For this project, students will conduct a 10–15 min interview with a community member to tell their story. Since the audio will feature at least two voices and tell an engaging story, students will first consider who their interview subject is and what messages they seek to convey through the interview. (Recommended design program: Adobe Audition).

Promotional Video: Students will create a 2–3 min informative video for their community members. Students will include a title screen at some point during the video and include any combination of found and/or original footage throughout the video. The video should demonstrate experimentation with multitrack video editing, accompanied by an edited voiceover and a polished soundtrack. (Recommended program: Adobe Premiere Pro).

Website Portfolio: Toward the end of the semester, students are expected to create a website portfolio that showcases all of the major assignments throughout the semester, including the empathetic audience awareness, advocacy poster, interview podcast, and promotional video. (Recommended program: Adobe Portfolio).

Calendar

Weeks	Objectives	Readings/tutorials	Due dates
1	• Overview of design thinking procedure and the rhetorical situations of multimodal composition • Critique of Sample Designs and Visual Arguments	Editor's Introduction Chapter 4	
2	• Introduce the Advocacy Poster assignment; Meeting community members to learn about their needs • Explore how to emphasize with community members and define the design problem	Chapters 1–2	
3	• Introduce design thinking principles and composition techniques for Advocacy Posters • Learn how to use Adobe Photoshop to engage in photo editing and poster design	LinkedIn Learning "Adobe Photoshop Essential Training" Tutorial	Empathetic Audience Analysis
4	• Have in-class peer reviews on the poster projects • Present rough posters to the community members	Chapters 3–4	Rough Poster Design Due
5	• Revise and iterate the poster design based on community members' feedback • Introduce the Interview Podcast Project	Chapters 5–6	
6	• Develop interview topics to amplify the voices of community members • Develop a list of interview questions and invite community partners to join the interview	Chapters 7–8	Final Poster Design Due
7	• Learn how to use Adobe Audition for editing podcasts • Workshop on removing background noise and adding sound effects	LinkedIn Learning "Adobe Audition Essential Training" Tutorial	

(Continued)

Weeks	Objectives	Readings/tutorials	Due dates
8	• Have in-class peer reviews on the podcast projects • Present rough podcasts to the community members	Chapters 9–10	Rough Podcast Due
9	• Revise and iterate the interview podcast based on community members' feedback • Introduction to Advocacy Video Project	Chapters 10–11	
10	• Learn how to create a storyboard for the promotional video project. • Gather digital assets, including license-free footages and music files online	Chapters 12–13	Final Podcast Due
11	• Learn how to use Adobe Premiere Pro for editing videos • Workshop on editing and remixing promotional videos with Adobe Premiere Pro	LinkedIn Learning "Adobe Premiere Pro Essential Training" Tutorial	
12	• Have in-class peer reviews on the video projects • Present rough video to the community members	Chapters 14–15	Rough Video Due
13	• Revise and iterate the promotional video based on community members' feedback • Introduce the website portfolio project	Chapters 16–17	
14	• Learn how to use Adobe Portfolio to design a website • Workshop on designing a final website portfolio that showcases the poster, interview, and video projects with the use of Adobe Portfolio	LinkedIn Learning "Building a Portfolio with Adobe Portfolio" Tutorial	Final Video Due
15	• Workshop on finalizing the website design • Individual conferences on finalizing the website portfolio	Editors' Outro	Website Portfolio Presentation
16	• Student Presentations of the website portfolios • Reflections on design thinking, multimodal composition, and socially-just design		Final Website Portfolio Due

INDEX

Note: *Italicized* pages refer to figures and page numbers followed by "n" refer to notes.